Frontiers of Space Risk

Natural Cosmic Hazards and Societal Challenges

Frontiers of Space Risk

Natural Cosmic Hazards and Societal Challenges

Edited by
Richard J. Wilman
Christopher J. Newman

CRC Press
Taylor & Francis Group
Boca Raton London New York

CRC Press is an imprint of the
Taylor & Francis Group, an **informa** business

CRC Press
Taylor & Francis Group
6000 Broken Sound Parkway NW, Suite 300
Boca Raton, FL 33487-2742

First issued in paperback 2020

ISBN-13: 978-0-367-57143-6 (pbk)
ISBN-13: 978-1-138-72638-3 (hbk)

Library of Congress Cataloging-in-Publication Data

Names: Wilman, Richard J., editor. | Newman, Christopher J., 1972- editor.
Title: Frontiers of space risk : natural cosmic hazards & societal challenges / edited by Richard J. Wilman, Christopher J. Newman.
Description: Boca Raton, FL : CRC Press, Taylor & Francis Group, [2018]
Identifiers: LCCN 2018003609 (print) | LCCN 2018027050 (ebook) | ISBN 9781315188294 (eBook General) | ISBN 9781351742672 (eBook - PDF) | ISBN 9781351742665 (eBook ePub) | ISBN 9781351742658 (eBook Mobipocket) | ISBN 9781138726383 | ISBN 9781138726383(hardback ;alk. paper) | ISBN 1138726389(hardback ;alk. paper)
Subjects: LCSH: Outer space--Exploration--Risk assessment. | Asteroids--Collisions with Earth--Environmental aspects. | Comets--Collisions with Earth--Environmental aspects. | Space environment. | Space law. | Space industrialization--Government policy.
Classification: LCC QB500.25 (ebook) | LCC QB500.25 .F76 2018 (print) | DDC 363.1/02--dc23
LC record available at https://lccn.loc.gov/2018003609

Visit the Taylor & Francis Web site at
http://www.taylorandfrancis.com

and the CRC Press Web site at
http://www.crcpress.com

Contents

Acknowledgments

THIS BOOK CAME ABOUT following a workshop on "Space Risk and Liability" held at Durham University, the United Kingdom, on June 29–30, 2015, coinciding with the inaugural *Asteroid Day* global awareness campaign.

The workshop was financed by Durham University's Institute of Hazard, Risk and Resilience, and brought together astronomers, space scientists, engineers, and technology consultants, alongside experts in space law and policy for two days of fruitful discussion. To further explore the workshop themes and related topics, we decided to assemble this interdisciplinary collection. We thank our commissioning editor Francesca McGowan for being receptive to the idea of blending such diverse themes within a single volume, and to Taylor & Francis/CRC's editorial staff Emily Wells and Rebecca Davies for their support and advice along the way. Finally, we thank all of our contributors for providing their chapters, which ultimately brought the book to fruition.

Richard J. Wilman
Christopher J. Newman

Editors

Richard J. Wilman earned a PhD in extragalactic astrophysics from Cambridge University, followed by a decade in astronomical research positions at universities in the United Kingdom, the Netherlands, and Australia. He has published over 50 papers on topics related to active Galactic nuclei, Galaxy clusters, and preparations for future astronomical surveys. Since 2012, he has taught physics and astrophysics at Durham University, the United Kingdom.

Christopher J. Newman is Professor of Space Law and Policy at Northumbria University, the United Kindgom. He was previously Reader in Law at the University of Sunderland. He has been active in the teaching and research of space law for a number of years and has worked with academics from other disciplines examining the legal, political, and ethical aspects of space exploration. He has made numerous appearances on British radio and television in relation to space law matters. He is a full member of the International Institute of Space Law and is also a member of the European Centre for Space Law.

Contributors

Mark E. Bailey
Armagh Observatory and Planetarium
College Hill
Armagh, Northern Ireland
United Kingdom

Thomas Cheney
Northumbria Law School
Northumbria University
Newcastle upon Tyne, United Kingdom

Camilla Colombo
Department of Aerospace Science and Technology
Politecnico di Milano
Milano, Italy

Pratika Dayal
Kapteyn Astronomical Institute
Groningen, The Netherlands

Mike Hapgood
RAL Space
STFC Rutherford Appleton Laboratory
Oxfordshire, United Kingdom

Christopher D. Johnson
Secure World Foundation
Washington, DC

Francesca Letizia
Astronautics Research Group
University of Southampton
Southampton, United Kingdom

Hugh Lewis
Astronautics Research Group
University of Southampton
Southampton, United Kingdom

Michael J. Listner
Space Law and Policy Solutions
East Rochester, New Hampshire

Ben Middleton
Department of Criminology, Law and Policing
Teesside University
Middlesbrough, United Kingdom

Christopher J. Newman
Northumbria Law School
Northumbria University
Newcastle upon Tyne, United Kingdom

Mirko Trisolini
Astronautics Research Group
University of Southampton
Southampton, United Kingdom

Mark Williamson
Space Technology Consultant
United Kingdom

Martin J. Ward
Department of Physics
Durham University
Durham, United Kingdom

Richard J. Wilman
Department of Physics
Durham University
Durham, United Kingdom

CHAPTER 1

Introduction

Christopher J. Newman and Richard J. Wilman

CONTENTS

1.1 WHY SPACE RISKS?

The notion of outer space as being a hostile and threatening environment is one which will be familiar to most people. Popular fiction is replete with tales of the dangers posed to the Earth from space (Roberts 2006). More prosaically, the growth in scientific understanding and human space activity since the start of the space race has led to an increased awareness of the dangers posed by an environment inherently hostile to organic human life. Equally, the conceptualization of space as a *frontier* has been identified as a dominant metaphor when considering space as a distinct environment (Billings 2004).* It is a metaphor, which works well as an overarching backdrop for the discussion of the risks, presenting both imagery of new possibilities and limitless opportunities while evoking a sense of physical

* The idea of space as a frontier has been subject to considerable academic inquiry. Although outside the focus of this discussion, the notion of frontier was identified by Billings (2004) as being drawn predominantly from American cultural history, with all of the assumptions and cultural implications that such a worldview brings. For further information, see also McDougall (1985).

danger and hazards. Such dangers lead to risks which can be both natural and human-made, tangible and invisible, and present challenges where the rule of law or its enforcement may be uncertain.

The aim of this book is to offer an introduction concerning how risks from space permeate and circumscribe humanity's increasingly diverse interactions with the space environment. This is underpinned by a widely drawn, multidisciplinary perspective, whereby scientific analysis and engineering solutions are discussed alongside the need for regulatory certainty and clarity in policy making. Fundamentally, as human activity in space increases, and our capacity to explore increases, the risks from such activity will also inevitably increase. As collective awareness of these risks increases, analysis and critique of potential solutions becomes ever more necessary to inform policy makers, embolden academics, and reassure the wider population. The immediate aim of the book is, therefore, to identify and discuss a range of risks coming from the space environment which might impact life on earth.

There are several reasons why such a discussion of risks to the Earth from space is of value. First, these risks that we identify herein will provide an alternate lens through which space activity and the study of space can be viewed and this will be of interest to a wide range of readers. Also, by appreciating the risks facing humanity, policy makers, engineers, and space professionals can decide upon the appropriate resource to dedicate to mitigating these risks. In some cases, where the risk is small, this might mean further resources with which to better understand the phenomenon. In other cases where the risk of occurrence is more likely, it might be that the need for capital expenditure on infrastructure is more pressing. Either way, by viewing risks from space holistically, we hope to provide increased awareness and inform public discourse on these areas. As Bostrom and Ćirković (2008) have already noted in respect of risk, "a broader view allows us to gain perspective and can thereby help us to set wiser priorities."

1.2 SPACE RISKS: FRAMING THE DEBATE

This first chapter will perform several functions. First, as already outlined, it will introduce and explain the value of studying the risks facing humanity from the space environment. In so doing, it is not intended to produce a definitive list of every risk faced by humans. Others have already successfully achieved this in a broader context (see, e.g., Bostrom and Ćirković 2008). Rather, it will reflect upon the nature of risks originating from various sources and how these risks can be categorized. Broadly speaking, these risks are categorized as being either natural cosmic hazards threatening life on Earth or risks that result from human activity in space. In the remainder of this chapter, we will explore these two themes and introduce each of the chapters in turn, to set them within the context of the book as a whole. The chapter will conclude with some final reflections and opportunities for further study.

All of the chapters in the book are written by contributors with a wide range of academic and practical expertise in various aspects of the space environment. Indeed, as can be seen, the contributors to the book are drawn not only from different disciplines, but from different academic traditions and backgrounds. One of the great strengths of such a multidisciplinary approach is the way in which it provides different perspectives on the same problems. Such an approach may lead to different but complementary solutions to the risks posed to humanity on Earth.

Any attempts to delineate between different categories of risks will be inherently arbitrary, nonetheless, there is a need for boundaries to be drawn for the inquiry through which the risks can be viewed and analyzed. It should be noted at the outset that this book purposefully limits itself to the risk to humans on Earth. While there are undoubtedly risks to humans in colonizing other planets and engaging on deep space missions of exploration, these are not likely to be serious ventures for many decades. As such, there are too many variables to make any concrete predictions as to the precise nature of these risks. Similarly, the book will not seek to delve too deeply into the theory of risk management and mitigation. There are numerous excellent specific discussions on this area (Macauley 2005) and consideration is, instead, focused on broader thematic inquiries into identifying the risks from space from a scientific and governance perspective.

1.3 THE CATEGORIES OF SPACE RISK

Having established the initial focus, the book operates from the basic precept that the threats from space to humans on Earth can be broken down into two broadly drawn themes. The first category of risk considered is that which originates from space and threatens all life on Earth. This encompasses the natural hazards arising from our cosmic habitat within the Solar System and the wider Milky Way Galaxy. These risks can be categorized as being naturally occurring (in so far as they are not the product of human activity) although they have the potential to impact both our natural environment and the critical national and global technological infrastructures upon which much of the fabric of the global society depends.

Such dangers are, of course, not limited to those posed by nature. The second broad area of risk comes, therefore, from human activity in space and the societal challenges such activity poses. This theme will encompass technological malfunctions and accidents affecting space hardware and infrastructure. This includes collisions in space due to the proliferation of orbital debris. Risks also stem from the deliberate misuse of space-based assets for the purposes of terrorism or state-based aggression. Such human activity requires an understanding of the attendant risks, and an appropriate mixture of engineering, financial, law, and policy solutions to mitigate them where possible or to maximize resilience.

As stated in Section 1.2 above, we are not primarily interested in technical, practice-based definitions of risk as they might apply to specific industries. We are not seeking to identify the risks purely in terms of scope (the number of those affected) and intensity (the impact of any risk). It is true that should some of the risks identified in the book come to pass then they will undoubtedly prove to be extinction events for life on Earth. Similarly, we are not interested in the precise quantification of risks, in the sense of identifying only those that fall within a certain range of probability; indeed, some of the risks identified may be considered highly likely, while for others the probability of occurrence is so small as to be negligible on timescales of human interest. All of these would provide unnecessary limitations on what is intended to be an exploration of a wide range of ideas.

1.4 THEMES WITHIN THE BOOK

As stated above, we did not impose a series of preconceived limitations based around qualitative categories of risk for authors to follow. Rather, we allowed those with the expertise in specific areas to identify the areas of risk and explore them independently, allowing each discrete chapter to reflect the vibrancy of contemporary research in that area. We have codified the contributions to this edited collection under the two broad themes of (1) natural risks, cosmic hazards and (2) human risks and societal challenges. In this section, we will highlight the key contributions and contents of each chapter within each theme.

1.4.1 Natural Risks and Cosmic Hazards

The start of the discussion on hazards emanating from the cosmic environment begins by considering the risk from asteroids and comets.

It is this risk with which humanity arguably has the longest and deepest association, at least on a cultural level. There is now a widespread public awareness of this hazard, perhaps due to the "impact" in the popular consciousness of the global mass extinction event that led to the demise of the dinosaurs 65 million years ago. More recently, a notable historical impact event occurred at Tunguska, Siberia, on June 30, 1908, in which the explosion of a ~100-meter-sized asteroid or cometary nucleus flattened some 2000 square kilometers of remote forest. The event is often cited as the benchmark for an impact with the capability to destroy a large metropolitan area, and is marked annually by the *Asteroid Day* global awareness campaign.* The threat was vividly demonstrated as recently as 2013 in the Russian city of Chelyabinsk, when the explosive atmospheric disintegration of a ~20-meter asteroid caused extensive property damage and personal injury, mainly due to flying glass shattered by the blast wave.

The spectacular appearance of bright comets has transfixed societies for millennia; our forebears considered them as harbingers of doom. The creation of artifacts from ancient and prehistoric civilizations, including some rock art and megalithic monuments, have been directly linked to episodes of elevated cometary activity (see e.g., Sweatman and Tsikritsis 2017), suggesting—as we have only recently begun to appreciate from astrophysical studies—that activity within the Solar System can evolve on relatively short timescales (see e.g., Napier 2010).

In Chapter 2, Mark Bailey provides a contemporary analysis of the astronomical understanding of the asteroid and cometary populations of the Solar System and the impact hazard they pose. He traces the growth in our knowledge of these *Near-Earth Objects* (NEOs), including basic observational properties such as their size distribution, alongside astrophysical theories for their origin. He explores the history of bombardment of the Earth, the effects of impacts by bodies of different sizes, and the risk they pose at present. By analogy with other low-frequency/high-impact geohazards such as tsunamis, volcanic eruptions, earthquakes, or large-scale floods he presents an insurance-based or actuarial approach to evaluating the average annual cost of NEO impacts and possible mitigation strategies.

Continuing with a Solar System focus, the second contribution on the theme of natural cosmic hazards considers the role of the Sun and its space weather impacts. Just as our ancestors may have observed

---

* https://asteroidday.org/.

disintegrating comets with a preternatural concern, the northern lights (*aurora borealis* and their southern counterpart the *aurora australis*) may have evoked a similar combination of fear and wonder among populations at high latitudes. For early societies, however, the impact of the typical auroral display was limited to the visual and aesthetic. It was only in the nineteenth century, with the development of electrical technology, that disruptive impacts began to be directly felt and the link with solar activity established. The first such instance occurred during the geomagnetic storm of early September 1859, when intercontinental telegraphy services were subject to anomalous behavior and disruption due to induced currents. The accompanying solar storm, dubbed the "Carrington Event" after Richard Carrington who recorded it (Carrington 1859), remains one of the strongest on record and continues to serve as the benchmark for space weather risk assessments.

With the dependence on electrical and electronic technologies in modern society, our vulnerability to space weather is now much higher. There are a wide range of potential impacts to technological infrastructure on the ground and in space, including, most notably, power grids, navigation, and communication systems. In Chapter 3, Mike Hapgood explains how we can track and forecast the impact of space weather and improve societal resilience to its worst effects. Such is the extent of the potential vulnerabilities that the issue has risen up the policy agenda and is now incorporated in the UK National Risk Assessment, alongside other natural and man-made disaster scenarios. As Hapgood concludes, planning for its effects must also feature in the development of new technologies, such as driverless cars, where the impact on digital devices and navigation systems could have lethal consequences if suitable safeguards and mitigation approaches are not in place.

Chapter 4 completes our coverage of natural cosmic hazards by casting the net beyond the confines of the Solar System and out into the Milky Way Galaxy and the Universe beyond. While Galactic hazards such as gamma-ray bursts, supernovae, and accreting black holes subjectively pose an entirely negligible risk to contemporary human civilization, at least in the sense that the probability of a dangerous event is extremely low on human timescales, they may have impacted the terrestrial environment over the course of the billions of years of Earth's history and have the potential to do so over the next billion years while the Sun permits the Earth to remain habitable. Astrophysicists Richard Wilman, Pratika Dayal, and Martin Ward review the science underlying the "threat" posed by these and other

Galactic hazards, and assess how they may have shaped the habitability of life-bearing planets in our Galaxy and beyond on cosmological timescales. With the explosion in recent decades in the number of known planets around other stars (so-called "exoplanets"), and the prospect that some may harbor soon-to-be-detectable life, they also touch upon some arguably more urgent and certainly more controversial risks surrounding our interaction with any extraterrestrial intelligence.

Their discussion invites a wider philosophical inquiry into the potential range of responses to the discovery of any signals from intelligent life elsewhere in the Galaxy. There is also a question as to the wisdom or otherwise of proactively sending targeted messages to other potentially life-bearing exoplanet systems. Aside from the great technical and sociological challenges associated with these tasks, they bring to light a number of societal considerations. With continuing rapid advances in the study of exoplanets such tensions may become more acute, especially if and when the first signs of life are discovered in exoplanet atmospheres, perhaps within the next decade.

In Chapter 5, we continue to explore the risks posed by physical hazards but move our attention away from those of the natural variety to consider those stemming from man-made space activity in Earth orbit. Camilla Colombo and collaborators examine the risks posed by space debris, both in orbit and on the ground. Earth orbit is littered by a growing population of space debris, deriving from the breakup of operational or defunct satellites and launch vehicles, a problem which will only be exacerbated by current trends in satellite miniaturization (specifically with the development of Cubesats and other smaller satellite platforms) and the proliferation of commercial launch providers.

Their discussion begins with an overview of the current debris population and its long-term evolution, before presenting an evaluation of the in-orbit collision risk and a detailed exposition of the techniques used to evaluate the threat to human populations and infrastructure on the Earth's surface from debris re-entering the atmosphere. Thereafter, they review the technical aspects of the various international measures and guidelines established to mitigate the space debris problem. This includes evaluation of collision avoidance and protection measures and mechanisms for Post Mission Disposal. The chapter ends with an illustrative discussion of so-called "passive end-of-life disposal" methods, which exploit natural orbital perturbations or solar radiation pressure and atmospheric drag.

1.4.2 Human Risks and Societal Challenges

The second part of the book shifts the focus away from risks and hazards which occur naturally or can be understood by scientific analysis. Instead, there is an examination of the way in which human interaction with the space environment has led to a series of risks all of which could have serious implications for human life on Earth. This section develops the problems of managing human-created debris, the threat from terrorism, the ever-present role of the military, and then finally considers the challenges to the Earth posed by a nascent space mining industry. Such considerations will be centered principally, although not exclusively, around the legal and policy-based solutions.

Although this part of the book will look at the overarching governance framework rather than focusing on the minutiae of space law,* there will be considerable discussion of the international law framework governing space activity. The Outer Space Treaty 1967† (OST) is the most significant element of international space law and is a treaty that was signed before humans had landed on the Moon. The legal framework surrounding the international governance of space has its roots, therefore, in the Cold War. As Gabrynowicz (2004) and Blount (2012) both identify, the primary purpose behind the drafting of the OST was to ensure security for the two cold war power blocs. It has been subject to significant criticism, including observations made within this book, about its relevance and suitability for the modern, multisectored space environment. Despite these criticisms, it remains the first port of call for any inquiry into the legality of space activity (see, for further discussion Lyall and Larsen 2009).

Initially, however, we start with an investigation of how resilience against the effects of space debris, space weather, and other stresses of operating in the space environment has been incorporated into the design and manufacture of satellites over many decades. In Chapter 6, space technology consultant Mark Williamson first explores the technical, or engineering approach, to the mitigation of such risks. Thereafter, he discusses how they can be mitigated financially, in the form of space insurance. In so doing, he

---

* There are a considerable number of "soft law" agreements and guidelines, which comprise the wider governance framework for space activities. See Chinkin (2000) and Goh (2008) for details.

† Treaty on Principles Governing the Activities of States in the Exploration and Use of Outer Space, Including the Moon and Other Celestial Bodies: 610 UNTS 205; 1968 UKTS 10, Cmnd. 3519; 18 UST 2410, TIAS 6347; (1967) 6 ILM 386; (1967) 61 AJIL 644. It was adopted by the General Assembly of the United Nations on December 19, 1966 and opened for signature on January 27, 1967 in London. It entered force on October 10, 1967.

draws on more than 30 years of his own personal experience in technical consultancy to space insurance brokers. He covers the history of the industry—dating back over 50 years, almost to the dawn of the Space Age—the nature of the market and its capacity, the types of coverage available, and the process of arranging cover and calculating premium rates and losses. He concludes with a look to the future of the industry, arguing that the demand for such cover will likely only increase as developments such as commercial space tourism and resource exploitation reach fruition.

In Chapter 7, Christopher Johnson of the Secure World Foundation revisits the question of human-made space debris and in doing so, links the technical coverage of the first theme with the law and policy analysis of the societal challenges. He explores notions of "Space Sustainability" in Earth orbit and considers what action needs to be taken to ensure that the space environment remains capable of meeting the needs of the present day and those of future generations. This discussion encompasses the space debris problem, complementing the technical coverage of Chapter 5 discussing the issue of nonbinding, debris mitigation guidelines (for further details see Hobe and Mey 2009). There is also consideration of potential problems such as radio frequency interference and concerns that beneficial space technology may not reach fruition due to political considerations which jeopardize its development. In doing this he studies emerging tensions in the Outer Space Treaty and their implications for the development of planetary defense mechanisms, resource exploitation, and the possibility of conflict in space.

Whereas the focus of the book to this point has been on unintended risks, it now shifts in Chapter 8 to intentional threats designed to cause significant disruption to space activity and the infrastructure on Earth. Ben Middleton brings his expertise as a researcher in terrorism to explore the risks posed to space activity from terrorism. The discussion identifies terrorism as a quintessentially human phenomenon and warns that the risk of terrorist activity expanding out into space must not be ignored. Middleton identifies that the most serious short-term threat comes from targeted activity against satellites and describes a number of scenarios where state and private terrorist actors could cause serious damage and disruption to the space infrastructure. The chapter ends with a call for bespoke policies and capital expenditure to protect vital space assets from being targeted by rogue states and terrorist groups.

Acknowledging that terrorist activity in space is likely to emanate from rogue states provides a useful segue into the next element of discussion of the risks in space from human activity. In Chapter 9, space attorney and founder and principal of Space Law and Policy Solutions, Michael Listner, illustrates the way in which human space activity has, largely, been the product of military activity. In a provocative and challenging discussion he maintains that in spite of the peaceful rhetoric embraced by academic commentators and various international institutions, most notably the United Nations, the attractiveness of space-based military assets and the development of advanced weaponry means that space remains a key theatre of military activity.

Listner argues that the current legal architecture for the governance of space activity is archaic and needs to evolve to more effectively manage the challenges and risks posed by military activity. This discussion provides an important counterpoint to the other contributions on policy and law. It reflects the realpolitik of modern geopolitics and illustrates the crossroads at which space governance finds itself. The certainties of the cold war and the consensus regarding the way in which space should be governed and developed have been eroded. The Outer Space Treaty and its cadre of related treaties still form the bedrock of normative behavior in space, yet the risk of instability and conflict is a genuine concern should this international consensus break down.

It is apposite that a significant risk to the international consensus is identified in the final chapter. Space law academics Thomas Cheney and Christopher Newman examine the way in which the nascent space mining industry could threaten the fabric of the international cooperation, which has underpinned space governance since the start of human activity. This chapter outlines that the current legal framework prohibits the appropriation of the Moon and other celestial bodies. A strict reading of Article II of the Outer Space Treaty would seem to suggest that states or private companies would not be able to lawfully possess any minerals mined from celestial bodies and even the very act of mining them could fall foul of international space law.

The lack of a compliant framework for the mining and trading of space resources risks the stability of the international governance framework underpinning human space activity. While the technological infrastructure of future mining ventures is at present unclear there is sufficient interest from private sector companies (particularly in the United States) to indicate that the legal problems of ownership will need to be addressed sooner rather than later. Both the United States and Luxembourg have already acted unilaterally to recognize mining as an aspect of the usage of space (as permitted under

Article I of the 1967 Treaty) rather than an attempt to "own" the celestial body. Cheney and Newman navigate through contemporary academic debate on this area and pose solutions based on a pragmatic adaption of existing international law. It is suggested that, given current geopolitical conditions, a new overarching space governance treaty is unlikely. The chapter finishes on an optimistic note, however, suggesting that the best way to ameliorate the risks from increased competition in space is by building on currently accepted norms and developing consensus on this area.

1.5 CONCLUSIONS AND OPPORTUNITIES

The scope and intensity of risks outlined in this book range from potential extinction events through to disruption of the current geopolitical governance arrangements for space activity (which may indeed be welcomed by some). The probability of these risks occurring also varies wildly—from the statistically insignificant through to the highly likely. What all of these risks have in common, however, is that they all occur in areas that would benefit from further academic study. While humanity may not currently be able to prevent the impacts of many of the natural cosmic hazards, this is not an immutable state of affairs and there is a clear advantage in achieving a deeper understanding of them.

The risks to Earth from human space activity may well have a higher probability of occurrence; while they are unlikely to result in the extinction of our species, they could have a significant impact upon our society. The "Kessler Syndrome" poses a clear risk to our ability to remain a spacefaring civilization and that threatens all space-based applications, communications, and GPS. The risk to life from terrorism or military activity and political instability may appear mundane in contrast to the natural cosmic hazards but their impact is no less worrying. Again, innovative legal and policy solutions are required to ameliorate the dangers from such risks. Going forward, it is suggested that we need clear and thoughtful leadership to ensure the implementation of broadly accepted governance frameworks.

As stated earlier on, it was a conscious decision not to assess the risks to humans *in space*. There has been some significant work already undertaken on this, such as Harrison (2001) on the risks to humans should we engage in long distance spaceflight and the work of Butler (2006) and Robinson (2006) on the issues facing human settlers on other worlds with respect to planetary contamination. These are all areas worthy of further study in a more suitable project and would undoubtedly enhance the understanding and appreciation of the risks to humans as space activity continues to develop.

A further, and wholly cognate, area of study that would draw on discussions which have been started by others, such as Dick (2015), center around the risk to humans from the discovery of extra-terrestrial life. The philosophical and cultural implications alone present a risk to the current fabric of society and there are significant questions which could be addressed in the context of both naturally occurring hazards from the discovery of life and the associated societal challenges.

Fundamentally, it is hoped that this book will stimulate a critical and enduring discourse concerning the nature of the risks posed by the space environment and the response to the societal challenges. This will allow policy makers, governments, and those with institutional responsibility to foster a clear and proportionate approach to the allocation of resources appropriate for managing these risks.

Perhaps the single most important contribution this book can make is by illustrating the utility of a truly cross-disciplinary approach to the study of space risks. Only by collaborating outside of the usual silos of activity will truly enduring solutions to managing the risks posed from the frontier of space be found. If this book in some way contributes to this process, then we will have realized our ambitions for the project.

REFERENCES

Billings L., 2004. How shall we live in space? Culture, law and ethics in spacefaring society. *Space Policy*, 22(4), 249–255.

Blount P., 2012. Renovating space: The future of international space law. *Denv J Int'l L & Poly*, 40(1–3), 515.

Bostrom N. and Ćirković M.M. (eds.), 2008. *Global Catastrophic Risks*. Oxford: Oxford University Press.

Butler J., 2006. Unearthly microbes and the laws designed to resist them. *Ga L Rev*, 41, 1355.

Carrington R.C., 1859. Description of a singular appearance seen in the sun on september 1. *Mon Not R Astron Soc*, 20, 13–15.

Chinkin C., 2000. Normative development in the international legal system. In: D. Shelton, ed., *Commitment and Compliance: The Role of Non-Binding Norms in the International Legal System*, 1st ed. Oxford: Oxford University Press, pp. 21–42.

Dick S.J. (ed.), 2015. *The Impact of Discovering Life Beyond Earth*. Cambridge: Cambridge University Press.

Gabrynowicz J., 2004. Space law: Its Cold War origins and challenges in the era of globalization. *Suffolk UL Rev*, 37, 1041–1066.

Goh G.M., 2008. Softly, Softly Catchee Monkey: Informalism and the Quiet Development of International Space Law. *87 Neb L Rev* 3 725–746.

Harrison A., 2001. *Spacefaring: The Human Dimension*. Berkley: University of California Press.

Hobe S. and Mey J., 2009. UN space debris mitigation guidelines/Die UN Richtlinien zur Verhutung von Weltraumtrummern/Lignes Directrices Relatives a la Reduction des Debris Spatiaux. *ZLW*, 58, 388–403.

Lyall F. and Larsen P., 2009. *Space Law*, 1st ed. Farnham, Surrey, England: Ashgate.

Macauley M., 2005. Flying in the face of uncertainty: Human risk in space activities. *Chi J Int'l L*, 6, 131–148.

McDougall W.A., 1985. *The Heavens and the Earth: A Political History of the Space Age*. New York: Basic Books.

Napier W.M., 2010. Palaeolithic extinctions and the Taurid Complex. *Mon Not R Astron Soc*, 405, 1901–1906.

Roberts A., 2006. *The History of Science Fiction*. London: Palgrave Macmillan.

Robinson G., 2006. Forward contamination of interstitial space and celestial bodies: Risk reduction, cultural objectives, and the law/Zur Kontamination des Weltraums: Risikobeschrankung, Kulturelle und Rechtliche Fragen/La Contamination de l'Espace Extra-Atmospherique: Reduction des Risques, Questions Culturelles et Juridiques. *ZLW*, 55, 380.

Sweatman M.B. and Tsikritsis D., 2017. Decoding Göbekli Tepe with archaeoastronomy: What does the fox say? *Mediterr Archeol Archaeometry*, 17, 223–250.

CHAPTER 2

Asteroid and Cometary Impact Hazards

Mark E. Bailey

CONTENTS

2.1 INTRODUCTION

For the first time in the c.4000 million-year history of life on Earth, a species—namely we human beings—has developed, which broadly understands Earth's place in space and the key factors that drive the development and evolution of life on Earth. Within just the last 50 years, humanity has come to realize that, contrary to the image popularized by Sagan's "Pale Blue Dot," the Earth does not hang isolated in space lit by a constantly shining Sun, but rather is a dynamic system interacting with and open to a constantly changing near-Earth celestial environment, and illuminated by a time-variable Sun. If extraterrestrial effects were tiny, we could be forgiven for ignoring them; but they are not. Rare impacts on the Earth by asteroids and comets, comprising the larger elements of the so-called Near-Earth Object (NEO) population, have driven major changes in the evolution of life on Earth over geological timescales and are key to the long-term future of the human race. Having recognized this risk, humans living in the twenty-first century are now faced with a unique responsibility to understand the risks and opportunities posed by Earth's near-space celestial environment and to respond appropriately. This chapter provides an overview of the hazards, as currently understood, posed by asteroids and comets in the Solar System.

2.2 BROAD VIEW

We are living through a unique period in Earth's history. From the earliest beginnings of life on Earth, less than a few hundred million years after the late heavy bombardment some 4000 million years ago, life has had a continuous toehold on Earth. For the first nearly 3000 million years of this enormous timescale, life's self-replicating organisms were fundamentally microscopic in scale and it was not until approximately 700–600 million years ago, leading up to the Cambrian Period (around 541 million years ago) and the beginning of the Phanerozoic Eon, that we begin to see abundant evidence for macroscopic fossil life. The fossil record presents us with an almost unimaginable range of different types of organisms, all of which appear destined ultimately to become extinct after timescales ranging from less than a million years to typically ten million years, or in some cases as long as several hundred million years; but in all cases on a timescale much less than the age of the Earth.

The story of the development of *intelligent* life on Earth is no less remarkable. Whereas most people would agree, to take a trivial example, that cats and dogs, or porpoises and whales, show evidence of intelligence in the way they interact with the world around them, it is not yet widely accepted that these kinds of creatures are self-aware—though to a lesser extent—in qualitatively the same way as humans.

Humanity, of course, is a proud species, with many members still inclined to see themselves as separate from others in the animal kingdom despite recognizing that humans are part of the evolutionary tree. Nevertheless, human beings do occupy a special place at the top of the tree when it comes to intelligence and in their ability to operate effectively in widely differing environments, as well as their power to affect the inanimate world around them including the terrestrial life forms with which they come into contact. Despite this advantage, there is no strong reason to believe that humanity will necessarily last longer than (say) the most highly successful groups in the fossil record, or that a new better adapted and even more capable entity might not one day evolve to replace it.

Apart from their innate intelligence and adaptability, humans have also—and this is a very recent development—learned much about planet Earth and its place in the cosmos. Within just the last few hundred years our collective astronomical knowledge has grown enormously: to encompass a coherent understanding of the first seconds of the entire 14 billion-year-old expanding Universe; the origin of the chemical elements of which we are all made; Earth's place in space; our Solar System's place in the Galaxy; and our Galaxy's place in the Universe. There are hundreds of billions of galaxies in the known Universe, each with tens or hundreds of billions of stars, of which many—perhaps the majority—will have planetary systems that might provide opportunities to possibly host exotic self-replicating organisms.

In the light of this broad understanding it seems inescapable that our Galaxy must be teeming with life. Some of these extraterrestrial life forms will be intelligent, as on Earth, and—like humans—possibly have the ability to recognize their place in space. Whether, like humans, they remain "Earth bound" or instead have grasped the opportunities of interplanetary and interstellar travel is currently unknown. However, because there is still no proof that extraterrestrials have yet visited Earth the average lifetime of technologically capable civilizations must be very short, or perhaps such species may prefer to stay at home; or possibly the undoubted difficulties of interstellar travel have proved insurmountable. Alternatively, perhaps we should get used to the idea that the development of a technologically adept species with the ability to leave its home planet is an extraordinarily rare occurrence in the evolution of life, and human beings are not just unique but possibly alone in the Galaxy?

Either way, we are inexorably drawn to conclude that life on Earth is precious. With humanity's growing understanding of the fragility of the various habitable zones of our planet, and their frailty on the cosmic stage, it behooves us to protect and preserve where possible the only place in the Universe where we are certain that intelligent life can exist.

In short, twenty-first-century humans—the first life-forms on Earth with the knowledge and power to act with space awareness—find themselves with a unique responsibility to understand better the risks and opportunities presented by our knowledge of Earth's variable near-space environment and to take appropriate action where possible to protect the species against existential risks. This view of humanity's place in the natural world is similar to that held by the philosopher and historian of science Toulmin (1982), who wrote: "Human Beings are the beneficiaries of history ... our fate within this historical scheme depends ... on the adaptiveness of our behavior [and] ... on the use of our intelligence in dealing with our place in Nature." We conclude that we should take action where necessary to protect the species and, where possible, other life forms on Earth as well. It is against this backdrop we now consider the newly identified asteroid and cometary risks to civilization, to human life, and more generally to the long-term survival of life on Earth.

2.3 POPULATIONS AND EFFECTS OF NEAR-EARTH OBJECTS

As indicated by the name "Near-Earth Objects" (NEOs), the term includes any macroscopic object—meteoroid, asteroid, or comet—which can potentially pass close by or collide with the Earth.

The sizes of these objects span a very large range, from objects just a few centimeters across which produce an occasional bright fireball that burns up high in the Earth's atmosphere, to bodies tens of meters across, which may penetrate much deeper into the Earth's atmosphere producing an airburst with an explosive yield similar to that of a nuclear weapon. A recent example is the approximately 20-meter diameter Chelyabinsk impactor, which ran into the atmosphere around dawn on February 15, 2013 at a speed of nearly 20 kilometers per second, producing an airburst with an energy of the order of 0.4 MT (1 MT of TNT equivalent is 4.2×10^{15} J) at a height of about 30 kilometers. A similar—though much more powerful—explosion occurred on the morning of June 30, 1908. In this case, the approximately 60-meter diameter Tunguska impactor disintegrated at a height of around 6 km.

The Chelyabinsk event was recorded by dash-cams over a very wide area around the city of Chelyabinsk, Russia, and produced a shock wave which smashed windows injuring more than 1500 people on the ground. The larger Tunguska event, which is estimated to have had an energy in the range 3–15 MT, flattened approximately 2000 km^2 of Siberian forest and produced atmospheric phenomena such as bright nights which were seen thousands of kilometers away.

At this low end of the NEO size distribution the objects are thought mainly to arise as a result of collisional fragmentation and orbital evolution of larger asteroidal bodies in the main asteroid belt, a region of the Solar System lying between Mars and Jupiter approximately 2.1–3.3 times the distance (1 astronomical unit, AU) of the Earth from the Sun. Some, however, come from comets, and the examples of Chelyabinsk and Tunguska illustrate both sources.

For example, the Tunguska event—in part owing to its trajectory and the coincidence of its occurrence with the time of the daytime Taurid meteors—is thought by many astronomers to be a fragment of the exceptional short-period comet 2P/Encke (e.g., Asher and Steel 1998). Others, however, argue for an asteroidal source (e.g., Chyba et al. 1993). Whichever view is correct, the millimeter to centimeter sized meteoroids that produce the majority of visible meteors or shooting stars certainly come from comets. Logically, there must be an intermediate-mass cometary component of the NEO population (with sizes ranging between those of the known kilometer-sized cometary parent bodies of meteor showers and microscopic interplanetary dust grains) and such cometary fragments must occasionally run into Earth. At the small end of the NEO size distribution, objects with diameters (d) in the approximate range 1–100 m are produced by both comets and asteroids.

As we move toward larger bodies, the effects of impacts become much larger. Objects with sizes in the approximate range 100–1000 meters or more will (depending on their internal structure, density, and impact velocity with the Earth) mostly pass straight through the atmosphere before reaching the ground with speeds approaching their cosmic velocity (some tens of kilometers per second) relative to Earth. These impactors will produce craters with typical dimensions (depending on factors such as density, speed, nature of the object, and material they hit) of the order of 10–20 times the diameter of the hypersonic projectile. That is, they produce craters with diameters, D, ranging from several hundred meters up to 10 kilometers or more and cause mostly local or regional, rather than global, devastation. Examples would be the Henbury crater field (Australia) with craters ranging in diameter from ~7 to 180 meters; Meteor Crater (Arizona) with D ~ 1.3 km, and the Steinheim–Reis crater pair (Germany) with D ~ 4 km and 24 km, respectively, believed to have resulted from the impact of a binary asteroid some 14.5 million years ago.

The energies associated with such impacts are enormous (~100–100,000 MT), much larger than anything yet produced by humans. In addition to the crater, the blast wave produced by an impact on land, and other effects (e.g., earthquakes, wildfires), will produce significant regional devastation (e.g., Toon et al. 1994). Similarly, objects landing in the sea may "crater" the ocean and produce a tsunami, which can affect coastlines hundreds or thousands of kilometers away (Hills et al. 1994).

There is a critical size above which the effects of an object's impact become global rather than local. The diameter above which this happens is slightly uncertain but is probably at the lower end of the range 0.5 to 2 kilometers. For example, the approximately 20 subkilometer size fragments of the progenitor of comet D/1993 F2 (Shoemaker-Levy 9), which ran into Jupiter between July 16 and 22, 1994 produced long-lived, opaque "bruises" in the visible atmosphere of that planet with dimensions comparable to the entire Earth. Similarly, the vaporized and melted dust lofted into the impact plume from a half-kilometer size impactor on Earth will greatly exceed that from the largest observed volcanic explosions. These arguments suggest that a significant global climate downturn could be caused on Earth by impactors with diameters perhaps as small as 0.5 km.

Summary tables of the effects of NEO impacts versus size have been compiled by many authors. Table 2.1 is adapted from Tables I and II of the chapter by Morrison et al. (1994), in the well-known *Hazards* book (Gehrels 1994). In calculating the impact energy, $E = 0.5\ mV^2$, where m is the mass of the projectile and V its velocity, we have assumed that the impactor has a density of 2 g cm^{-3} (it would probably be less than half of this value for a low-density cometary fragment), and an impact speed of the order of 25 km s^{-1} (an asteroid impact could be slightly less than this, whereas comets may impact with speeds ranging up to 60 km s^{-1} or more). Throughout the NEO size range, from diameters less than 0.5 km to more than 10 km, both asteroids and comets contribute to the impact hazard.

As this table shows, NEO impacts can cause environmental damage over areas ranging up to those of countries, continents, and in extreme cases the entire Earth. Impacts by the largest members of the NEO population (with diameters of the order of 10 kilometers or more) are thought to be capable of causing a mass extinction of species, such as happened 65 million years ago with the extinction of the dinosaurs. Events such as these can change the course of evolution of life on Earth, and at their most extreme could lead to the extinction of all life on Earth.

TABLE 2.1 Effects of NEO Impacts on Earth versus Size

NEO Diameter d (km)	Impact Energy E (MT)	Type of Event	Crater Diameter D (km)	Consequences
<0.05	<10	Fireball, Airburst	N/A	Upper atmosphere detonation of stones and comets; only irons penetrate to surface. Limited local devastation.
0.05–0.3	10–2000	Tunguska-like events	1–6	Irons make craters (cf. Meteor Crater). Stones and comets produce airbursts. Land impacts destroy a large metropolitan area or a small country at the large end. Oceanic impacts produce tsunamis.
0.3–1	2000–100,000	Sub-global land impacts	6–30	Tsunamis reach ocean scales. Land impacts destroy an area the size of a medium-size country. Land impacts raise enough dust to affect climate and produce a global effect.
1–10	0.1–100 million	Global impact	30–200	Both land and ocean impacts raise dust, changing climate. Land impacts destroy an area the size of a large country. Impact ejecta are global, triggering widespread fires. Prolonged climate effects, global conflagration. Probable trigger for mass extinction of life.
>10	>100 million	Global impact	>200	Large mass extinction; large impacts threaten the survival of all advanced forms of life.

2.4 NUMBER OF NEAR-EARTH OBJECTS

Sixty years ago, the number of known NEOs was relatively small and the frequency of large-body impacts was thought to be relatively low; in consequence, the risk to civilization posed by comet or asteroid impacts was small enough to be ignored. The first main-belt asteroid, Ceres, was discovered in 1801 followed soon after by Pallas, Juno, and Vesta. However, it was to be nearly a further hundred years before the first Earth-approaching asteroid, Eros (which belongs to the Amor sub-class), was found in 1898 while the first Earth-crossing asteroid, Apollo, was discovered only in 1932. Based on the then available data, Whipple (1973) estimated that the total number of Apollo asteroids with diameters $d > 1$ km would be around 100.

Despite this suggestion that NEOs could probably be ignored the number of known objects on Earth-crossing orbits has since grown enormously, and this has led to a sea change in our assessment of the risk. Twenty years after Whipple's estimate, the projected number of near-Earth asteroids (NEAs) was thought to be about 10 times larger, in the approximate range 1000–2000 (Rabinowitz et al. 1994), while a more recent estimate (Rabinowitz et al. 2000) suggested a figure about half this.

Table 2.2, using data from the NASA JPL CNEOS website (2018), illustrates this rapid increase in the number of known NEOs. It should be emphasized that the increase in the number of discovered NEOs is not a *real* increase in the number of such objects, which has remained rather constant over this timescale, but rather an indication of the improvement in our knowledge of Earth's near-space environment. The declining rate of increase in the number of known NEOs with diameters greater than 1 km during the last decade now suggests a projected figure of the order of 1000 for the total number of such NEOs (e.g., Bottke et al. 2002).

TABLE 2.2 Growth in Knowledge of Near-Earth Objects

Start of Year	Number Near-Earth Asteroids	Number Near-Earth Comets	Number NEAs with Diameters >1 km
1970	27	41	24
1980	51	44	43
1990	133	48	88
2000	878	57	305
2010	6649	91	790
2015	11,951	100	862
2017	15,403	106	877
2018	17,459	107	885

2.5 FREQUENCY OF IMPACTS

There are broadly two independent ways to estimate the NEO collision rate. The first is simply to look upward, identify as full a catalogue as possible of NEOs versus size and orbital type, and work out statistically the probability that one of these objects will hit the Earth. Another approach is to look down, and seek ground truth as it were, by counting craters on the Earth versus size and age. This approach lends itself to counting craters on the Moon and also, although the detailed connections are harder to make, on other Solar System objects as well such as Mercury, Venus, Mars, the few asteroids visited by spacecraft, and the cratered surfaces of the satellites of the outer planets.

2.5.1 Looking Up

Astronomers classify NEOs into two main types—asteroids and comets (and their fragments)—and among the asteroids there are four broad orbital subdivisions: Apollo and Aten asteroids, and Amor and Atira (or "Inner Earth Orbit") asteroids. Of these, only the Apollos and Atens (together comprising about 60% of the known NEO population) currently have orbits that potentially allow a collision with Earth, but the Amor and Atira types are considered because their approach to Earth is sufficiently close to be included in the conventional lists of near-Earth asteroids. Over long periods of time, the orbits of NEOs can change, leading to evolution from one orbital class to another, bringing objects not currently Earth-crossing into orbits that are.

Apollo asteroids are defined as those with semimajor axes >1 AU (i.e., with orbital periods greater than that of Earth) and perihelion distances, q, less than Earth's current aphelion distance ($Q_E = 1.017$ AU) so they have a possibility of colliding with the Earth. Atens have semimajor axes a <1 AU and aphelion distances greater than Earth's current perihelion distance, $q_E = 0.983$ AU. Amors have orbits completely beyond that of the Earth, that is, perihelion distances greater than $Q_E = 1.017$ AU, but less than 1.3 AU, which is the conventional minimum perihelion distance for an object to be classified as "near-Earth." The numerically small group of Atira asteroids have orbits contained entirely within Earth's orbit, that is, they have aphelion distances less than Earth's perihelion distance, $q_E = 0.983$ AU.

Considering that there is a projected population of approximately 1000 NEOs with diameters greater than 1 km, with an average terrestrial impact probability of the order of 5×10^{-9} per observed object per year (e.g., Steel 1995), the estimated rate of impacts of such objects with the Earth will be of the order of 5×10^{-6} yr^{-1} corresponding to a mean interval between such impacts of the order of 200,000 years. The cumulative size distribution of NEOs in the relevant size range ($0.07 < d < 3.5$ km) is approximately a power law proportional to d^{-2} (e.g., Jeffers et al. 2001), so the mean interval between impacts of objects large enough to cause a global climate downturn ($d > 0.5$ km) is of the order of 50,000 years. For larger objects, that is, those with $d > 3.5$ km, the cumulative number of NEOs falls much faster than the above inverse square law (e.g., Morbidelli et al. 2002) and can be approximated by a rather steep power law with index 5.4. Comets, on the other hand, have no such sharp cutoff in their cumulative size distribution, at least not until one reaches diameters at least of the order of 30–50 km or larger.

2.5.2 Looking Down

This approach depends on identifying hypersonic impact craters on Earth, and dating them as closely as possible, and then using crater-diameter scaling relationships to infer the size, velocity, and nature of the projectile which would most likely have produced each crater. Grieve and Dence (1979) showed that the largest terrestrial craters, that is, those with diameters larger than approximately 20 km, for which the discovery statistics of surviving craters are nearly complete, have a cumulative size distribution roughly proportional to D^{-2} (see Ivanov et al. 2002 for a more comprehensive discussion). This suggests that although NEAs might dominate the cratering rate for small craters, the largest craters would seem more likely to be produced by rare cometary impacts (cf. Yeomans and Chamberlin 2013).

Following the review by Bailey (1991), the overall terrestrial cratering rate for craters with diameters larger than D can be estimated with some uncertainty to be given approximately by $N(>D) = N_{20} \times 10^{-6} (D_{20})^{-2}\ \mathrm{yr}^{-1}$, where $D_{20} = D/(20\ \mathrm{km})$ and the leading coefficient N_{20} is of the order of 3 ± 2. If kilometer-size asteroids produce craters with diameters of the order of 20 km, then the two approaches—looking up and looking down—agree with one another within the errors. However, considering the uncertainties underlying each approach a significant contribution from objects other than the currently observed near-Earth asteroid population cannot be ruled out.

2.6 GIANT COMETS

As noted above, the size distribution of comets—like that of asteroids in the main belt—extends upward to very large values, much greater than those of even the largest known near-Earth asteroids. The principal cometary reservoirs are first the Oort cloud, which extends more than 100,000 AU from the Sun, with some comets moving in orbits extending more than halfway to the nearest star; and second, the several distinct populations of trans-Neptunian objects (the classical Edgeworth–Kuiper belt, the Scattered disc, objects in low-order mean-motion orbital resonances with Neptune, and so on), which are a thousand times closer. Taken together, these cometary reservoirs are estimated to contain more than a hundred billion kilometer-size objects and very large numbers of objects with sizes greater than 50 or 100 kilometers across.

It turns out that some of these outer Solar System objects, for example Pluto (the first trans-Neptunian object to be discovered, in 1930, and the so-called "king" of the Edgeworth–Kuiper belt), have sizes ranging up to nearly planetary size and, like the main-belt asteroid Ceres, have recently been reclassified as dwarf planets. With this in mind, commentators have sometimes asked what would such an object look like and what would be its effects if such a large icy object was ever to be scattered by planetary perturbations into the inner Solar System. It is hard to avoid the conclusion that it would be terrifying and probably look cometary.

While no one is suggesting that such a large object is likely to fall into the inner Solar System, there are nevertheless on the order of 100,000 objects in the trans-Neptunian region with diameters greater than 100 km (cf. Trujillo et al. 2001). Some of these objects will be transferring into shorter period orbits of smaller perihelion distance, first into the region occupied by Centaurs (a class of icy asteroids or comets which can be defined as those with perihelion distances $5 < q < 28$ AU and semimajor axes less than 1000 AU; e.g., Emel'yanenko et al. 2013), and then evolving into short-period cometary orbits under the dynamical control of Jupiter. Once in such an orbit they would be expected to decay under the effects of solar heating or fragment into their constituent components (or be dynamically scattered back into the Centaur zone, or possibly ejected from the Solar System entirely to become denizens of interstellar space).

The first Centaur, namely Chiron, was found in 1977 with a diameter of approximately 200 km. It moves in a long-term unstable orbit, with a current orbital period of approximately 50 years and perihelion distance $q = 8.5$ AU close to that of Saturn. During a dynamical lifetime measured in millions of years, it is predicted to present itself as a typical Jupiter family short-period comet perhaps up to a dozen times within an overall dynamical lifetime of several million years before ejection from the Solar System (Hahn and Bailey 1990). In fact, there is an *a priori* probability of just 10% that Chiron has not previously been a short-period comet.

This highlights the fact that estimates of the average NEO population over very long timescales, for example to compare with the terrestrial cratering record over the Phanerozoic, must take account of objects such as Centaurs and Oort cloud comets. Although such objects may currently be moving in orbits far removed from near-Earth space they may nevertheless be inserted into NEO orbits for periods of time ranging from hundreds to tens of thousands of years as a result of ordinary chaotic dynamical evolution.

2.7 A GREATER ROLE FOR COMETS

Despite the attention given in recent years to the *observed* NEO population there has been a long-standing parallel debate in the literature since the late 1970s—after the first Centaur was discovered and when the asteroid hazard began to be more widely recognized, as to the importance of comets for a correct assessment of the terrestrial impact hazard and in particular for the long-term risk to civilization posed by giant comets. Work by a number of authors (e.g., Hoyle and Wickramasinghe 1978; Napier and Clube 1979; Clube and Napier 1982, 1984; Asher and Steel 1993; Hoyle 1993; Asher et al. 1994; Bailey et al. 1994; Bailey and Napier 1999; Napier 2015) has emphasized key differences between the predicted hazards to civilization posed by the "one-off" random impact of a kilometer-size asteroid, on average perhaps every 100,000 years, versus the far more complex picture of the injection—on a broadly similar recurrence timescale—of a giant comet from the outer Solar System and its subsequent decay in the region of the terrestrial planets.

Assuming an approximate inverse-square cumulative size distribution for comets, that is, one in which the number of objects with diameters greater than d is approximately proportional to d^{-2} (e.g., Snodgrass et al. 2011; Fernández and Sosa 2012; Bauer et al. 2013), a giant comet with $d > 50$ km should be injected into a short-period cometary orbit on average once every approximately 50,000 years. Such a comet undergoing normal evolution under the effects of solar heating and close planetary encounters would eject copious amounts of dust and larger meteoroids into the interplanetary complex and probably undergo hierarchical fragmentation into trails and streams of smaller objects, as appears to have happened in the case of the progenitor of the Kreutz Sun-grazing comet family (e.g., Marsden 1989; Öpik 1966).

Such a giant comet would present a coherent, time-variable impact hazard to the terrestrial planets. Depending on its detailed physical evolution and the nodal crossing times of the parent body's meteoroid streams, such a model provides a much more satisfactory astronomical framework to explain the complexity of the geological record around the time of the Cretaceous-Tertiary (K/T) boundary (e.g., Bailey et al. 1994 and references therein) than the alternative one-off asteroid impact. The nature of the hazard to civilization owing to the most recent giant comet injected into the inner Solar System, perhaps as recently as 10–20 thousand years ago, also has implications for the way we interpret a range of more recent puzzles in the geological record, that is, the prehistoric period leading up to the end of, and since, the last Ice Age (e.g., Steel et al. 1991; Asher et al. 1994; Bailey 1995; Napier et al. 2015).

2.8 RISK ANALYSIS

Evaluating the potential risk to civilization of asteroids and comets on Earth-crossing orbits presents a number of intractable problems compared with the assessment of ordinary risks with which we are more familiar. First, the science is still relatively poorly understood. Unlike much more frequent high-consequence natural disasters, such as those caused by storms, floods, earthquakes, or volcanoes we have no recent experience of the phenomenon. Estimates of the frequency of asteroid or comet impacts are rather uncertain and we still have only limited ways to calibrate their effects. Furthermore, when the mean recurrence interval between successive events of a particular type is measured in tens or hundreds of thousands of years, there is an almost irresistible temptation for people to ignore the objective risk—perhaps on the ground since it is very unlikely to crystallize during their own lifetime.

Second, the impact hazard is unique among all the potential threats to civilization in that the size of the risk, however calculated, is potentially unbounded. In short, it is not just the survival of a country, or modern civilization that is at stake, but possibly life on Earth itself. Yet, the impact hazard is amenable to mitigation. In the context of our modern understanding of Earth's place in space, if the potentially infinite risk can be mitigated, then logically it should be a priority to address.

Third, whereas the trajectory of a powerful storm can be predicted with some precision usually at most only hours or sometimes days in advance, and the timing of an earthquake or volcano much less accurately, the location and time of an asteroid or comet impact can be predicted with high accuracy possibly years or decades in advance, thus providing time for a variety of possible mitigation measures to be implemented. All that is required is to discover the offending comet or asteroid sufficiently early before impact to determine its orbit accurately.

A complicating factor in this case is the question of who owns the risk. Is it us, because we have identified it, or—perhaps more reasonably because it potentially involves a global threat—should all nations on Earth contribute to solving the problem, perhaps in proportion to a country's area or the size of its population or economy? This is a situation where the wealthiest nations have the most to lose. But there remains a possibility that no single group will regard the matter as of sufficient importance to command early attention and so the issue may be avoided and the risk, no matter how large, left untouched.

2.8.1 Actuarial Assessment

Leaving aside these complications, it is relatively straightforward to estimate the order of magnitude of the NEO risk in comparison with other rare high-consequence risks which governments routinely consider. We live in a society bombarded with concepts relating to risk, from the "risk" of a financial crash to that of a cyber-attack on our national infrastructure, to a statistically overdue global influenza pandemic. In every case, evaluation of the risk requires accurate knowledge of two parameters: (*i*) the chance, or more accurately the probability per unit time, of a given event happening; and (*ii*) the nature of the given event, precisely defined, and its consequences.

Of course, not every risk is negative in the sense of being undesirable, and risk takers are often among the most successful entrepreneurs: the value of a positive opportunity must also sometimes be thrown in the mix. In the case of the extinction of the dinosaurs, for example, it is arguable it was precisely this particular random event in Earth history that led to the circumstances facilitating our own existence 65 million years later. Good for humanity; bad for the dinosaurs! Positive benefits arising from the discovery of NEOs also include the possibility of exploiting a particular asteroid or comet for its rare metal or mineralogical content (e.g., Lewis 1997; see also Chapter 10), or using NEOs to discover more about the origin of the Solar System, or simply as stepping stones to help explore the planetary system.

From this discussion, it is evident that the quantitative evaluation of a particular risk can sometimes become highly subjective, more art than science, and informed by factors other than a supposedly objective assessment of the frequency of a particular event and its consequences. When considering a particular risk, we must also be aware of the danger of tunnel vision, the effect of a misleading scientific consensus masquerading as fact, or even of bias informed by social media and the reaction of society at large. Such complications may lead to an entirely unjustified response to particular types of event, possibly leading other risks to be overlooked and skewing the risk matrix in such a way that significant resources may be wasted in managing virtual threats down to levels out of all proportion with their true frequency and cost.

If we look dispassionately at these matters, it is clear that the risks dealt with most satisfactorily are those in which both the nature of the event and the frequency of its occurrence are well defined. Usually, this means that we have strong empirical grounds to support our figures, as for example in the familiar cases of home or travel insurance, or the frequency of deaths arising from road, rail, or air transport disasters. This motivates the development of an actuarial approach to risk, which although not universally accepted, has the merit in the present context of allowing intrinsically very different kinds of events to be compared one with another, including the exceptionally rare high-consequence effects associated with NEO impact hazards.

With an actuarial approach, we estimate the "cost" of any given event by multiplying the number of implied fatalities by an agreed numerical factor, say £1.5M for the United Kingdom, which represents the economic value of an average human life. Some see this as an emotionally cold approach to risk but it allows us to estimate the average cost to the United Kingdom per year of a typical Tunguska-like impact event and compare it, for example, with a much rarer but much larger impact of a kilometer-size asteroid.

Thus, the frequency of Tunguska-like events is of the order of one per century (e.g., Asher et al. 2005). This leads to a frequency of around 0.01 per year that an area of the order of 2000 km^2 somewhere on the Earth will be flattened by the blast wave arising from the impact of a sub-100 meter diameter meteoroid. The total area of the United Kingdom is about 0.05% of the area of the Earth and 2000 km^2 is approximately 1% of the area of the United Kingdom, which has a total population of approximately 70 million people. Assuming that everyone living within the approximately 2000 km^2 tree-fall area of the Tunguska-like event, comparable to the area within London's ring motorway, would be killed, the actuarial cost of Tunguska-like events for the UK alone would be of the order of £5M per year. Of course, this is a very rough estimate, which takes no account of details such as the concentration of people in cities across the United Kingdom or the size distribution of impactors, which means that geographically larger areas would be affected by larger bodies, albeit less often.

When we consider impacts by kilometer-size objects, that is, by objects large enough lead to a global climate downturn and to the death of perhaps a quarter of the world's human population (e.g., Chapman and Morrison 1994), a similar analysis leads to an impact frequency of about one per 200,000 years and (for the UK alone) the loss of a quarter of its population. In this case, the actuarial cost of the event (again, for the UK alone) is in the order of £130M per year, an order of magnitude greater than the average risk posed by smaller Tunguska-size impacts.

Such estimates can, of course, be refined (e.g., Canavan 1994) but the example illustrates a number of key points surrounding the asteroid and comet impact hazard to civilization. First, the small (albeit much more frequent) impacts present a negligible risk compared to the much larger (but very rare) events, which have a global reach. This raises the question of the ranking of risks and their mitigation, for example, how much money should be spent proportionately on discovering small NEOs compared to the rare large ones, and how much should be spent on mitigating their respective risks.

In partial answer to this question, it is evident (again, for the UK alone) that the actuarial, or insurance, cost is affordable in the sense that sums even as high as 10 million pounds per year could presumably be found if necessary to "defend" the United Kingdom against Tunguska-size impacts. Even figures ten times larger could in principle be found by governments in order to mitigate the risk posed by kilometer-size asteroids whose impacts would have a global reach. Whether such sums of money *should* be spent depends on the community's risk appetite, and perhaps also on a comparative assessment of the quantum of resource allocated every year to mitigating other high-consequence events which figure on the UK Government Risk Register and might on average lead to a similar or greater number of deaths per year (e.g., Bailey 2011). A full discussion of these issues was given by Holloway (1997) who concluded, "If NEOs were a business, they would not be allowed to operate."

Given that the largest NEOs present by far the greater risk to human civilization, it is fortunate that their total numbers (with diameters larger than a kilometer across) is only about 1000 and that they are the ones most easily found. The catalogue of known near-Earth asteroids with diameters greater than 1 km is currently thought to be at least 90% complete. Once complete it will be possible to retire the asteroid impact hazard and replace it with a binary answer to the question whether the Earth will be hit by such an asteroid within, for instance, the next 100 years. Nevertheless, despite the objectively much lower risk presented by the astronomically smaller bodies, it seems likely there will remain a pressure to discover more of the many tens or hundreds of thousands of smaller bodies—if only because it will be argued that we cannot live with the unexpected risk of a Tunguska-size event, which statistically (with a recurrence interval of the order of 100 years) is almost certain to occur before we are hit by a much larger body.

2.8.2 Other Considerations

It was Donald Rumsfeld, who in 2002 became for a short time the patron saint of risk analysis with his remark to the effect that what matters most are not the known knowns, or even the known unknowns, but rather the *unknown unknowns*. This highlights an important lesson for proponents of risk analysis; that is, how easy it is for us to be trapped by a kind of groupthink, or tunnel vision. When we think we understand a given situation, we can still find ourselves blindsided by a sudden change of circumstances or the appearance of a factor we may have subconsciously or deliberately chosen to ignore, or about which we literally knew nothing.

The analogous uncertainty in the case of the comet and asteroid impact hazard is the significance of the role played by comets, which are often ignored. In part, this is because their long periods and potentially large sizes mean that current determinations of the risk are subject to particularly large uncertainties. In part also, it is because of the uncertain effects of giant comets. Although giant comets do not pose an immediate threat, there remains a need to understand better the processes of their orbital evolution and the effects of the breakup of the most recent giant comet in a short-period orbit, which perhaps occurred as recently as 10,000 or 20,000 years ago, and to take account of its continuing effects.

The review by Napier et al. (2015) presents a range of arguments that Encke's comet and the associated Taurid Complex may have played an important role during this time-period, perhaps being responsible for the environmental downturn around 12,800 BP and a factor in subjecting the Earth to Tunguska-like bombardment events (at intervals of the order of 3000 years) when the trails of fragmenting cometary debris became node-crossing with the Earth. The evidence amassed by some scientists for a cosmic or even cometary cause for relatively recent terrestrial environmental disturbances is controversial, but a significant time-variable cometary contribution to the interplanetary complex cannot be ruled out. Indeed, the possibility raises very interesting questions for the way in which ancient societies during historic and proto-historic times might have regarded their near-space celestial environment, obsessed—as it appears—by changing astronomical phenomena and placing their gods in the sky (e.g., Clube and Napier 1982; Hoyle 1993; Bailey 1995). Crosscutting investigations of this kind seem certain to form the basis for future interdisciplinary research.

2.9 CONCLUSION

We are living through a unique period of Earth's history. For the first time in the history of life on Earth, a species has developed which understands Earth's place in space, and the fact that Earth is an open system in touch with its near-space celestial environment. The discovery of near-Earth asteroids and comets, and rare giant comets in particular, informs our present understanding that the survival of civilization—and even the long-term sustainability of life on Earth—is contingent upon random Solar System events. That is, dependent on rare collisions of kilometer-size and larger asteroids and comets with the Earth, and on the arrival (on timescales of the order of 100,000 years) of rare giant comets in short-period cometary orbits in the inner Solar System. The resulting stochastic and episodic extraterrestrial influences on the Earth lead to a range of rare, high-consequence effects on the terrestrial environment, which in turn present society with unusual difficulties so far as risk assessment is concerned.

The broad view provided by our modern understanding of Earth's place in space leads us to conclude that whereas life in one form or another must be extremely common among the tens of billions of habitable planets in the Galaxy, *intelligent* life with the capacity to grasp its place in the wider Universe and with a capability to travel beyond its home planet must be very rare. On Earth, this has happened only once in more than 4000 million years of evolution of life on this planet, and it is by no means certain that humanity will learn this important lesson of Earth history.

During the last approximately 50 years, humanity has developed a high level of space awareness. Within this same timescale, the work of astronomers around the globe has provided increasing evidence to support a new worldview: a paradigm shift arguably as important as that, 400 years ago, which consolidated the heliocentric revolution. The new worldview emphasizes the position of the Earth as an open—rather than a closed system, intimately interacting with its near-space celestial environment.

This chapter has focused on one important aspect of this interaction, namely rare but extremely high-consequence impacts of near-Earth asteroids and comets with the Earth, and the effects of streams of meteoroids and larger bodies, as well as copious amounts of dust injected into the interplanetary complex through the medium of giant comets.

We recognize that these extraterrestrial influences on the terrestrial environment play an occasional major role in the evolution of life on Earth and, on much shorter timescales, in creating stresses which may drive the development of civilization.

There is much work to be done before a full understanding of the existential threat to civilization posed by Earth-crossing asteroids, comets, and cometary debris is achieved. It is a truism that the first step to solving a problem is recognizing that there is one, and it is here that astronomy—arguably the queen of sciences, but rarely useful in a hard, economic sense—has really proved its worth. If the question is the risk to life and to human civilization as a result of Earth's interactions with its near-space environment, the words of the French physicist Jean Perrin (1870–1942) are particularly apt: "It is indeed a feeble light that reaches us from the starry sky. But what would human thought have achieved if we could not see the stars?" What too might humanity achieve by an even greater level of space awareness?

REFERENCES

Asher, D.J., Steel, D.I. 1993. Orbital evolution of the large outer solar system object 5145 Pholus. *Mon. Not. R. Astron. Soc.* 263: 179–190.

Asher, D.J., Steel, D.I. 1998. On the possible relation between the Tunguska bolide and comet Encke. *Planetary and Space Sci.* 46: 205–211.

Asher, D.J., Clube, S.V.M., Napier, W.M., Steel, D.I. 1994. Coherent catastrophism. *Vistas Astron.* 38: 1–27.

Asher, D.J., Bailey, M.E., Emel'yanenko, V.V., Napier, W.M. 2005. Earth in the cosmic shooting gallery. *Observatory* 125: 319–322.

Bailey, M.E. 1991. Comet craters versus asteroid craters. *Adv. Space Res.* 11: 43–60.

Bailey, M.E. 1995. Recent results in cometary astronomy: Implications for the ancient sky. *Vistas Astron.* 39: 647–671.

Bailey, M.E. 2011. Risk and natural catastrophes: The long view. In *Risk*, eds. L. Skinns, M. Scott, T. Cox, 131–158. Cambridge, UK: Cambridge University Press.

Bailey, M.E., Clube, S.V.M., Hahn, G., Napier, W.M., Valsecchi, G.B. 1994. Hazards due to giant comets: Climate and short-term catastrophism. In *Hazards Due to Comets & Asteroids*, ed. T. Gehrels, 479–533. Tucson, AZ: University of Arizona Press.

Bailey, M.E., Napier, W.M. 1999. The fluctuating population of Earth impactors. *J. Br. Interplanet. Soc.* 52: 185–194.

Bauer, J.M. et al. 2013. Centaurs and Scattered disk objects in the thermal infrared: Analysis of WISE/NEOWISE observations. *Astrophys. J.* 773: 22 (11pp).

Bottke, W.F., Cellino, A., Paolicchi, P., Binzel, R.P. 2002. *Asteroids III*. Tucson: University of Arizona Press.

Canavan, G.H. 1994. Cost and benefit of near-Earth object detection and interception. In *Hazards Due to Comets & Asteroids*, ed. T. Gehrels, 1157–1189. Tucson, AZ: University of Arizona Press.

Chapman, C.R., Morrison, D. 1994. Impacts on the Earth by asteroids and comets: Assessing the hazard. *Nature* 367: 33–40.

Chyba, C.F., Thomas, P.J., Zahnle, K.J. 1993. The 1908 Tunguska explosion: Atmospheric disruption of a stony asteroid. *Nature* 361: 40–44.

Clube, S.V.M., Napier, W.M. 1982. Spiral arms, comets and terrestrial catastrophism. *Quarterly Jl. Roy. Astron. Soc.* 23: 45–66.

Clube, S.V.M., Napier, W.M. 1984. The microstructure of terrestrial catastrophism. *Mon. Not. R. Astron. Soc.* 211: 953–968.

Emel'yanenko, V.V., Asher, D.J., Bailey, M.E. 2013. A model for the common origin of Jupiter family and Halley type comets. *Earth, Moon, Planets* 110: 105–130.

Fernández, J.A., Sosa, A. 2012. Magnitude and size distribution of long-period comets in Earth-crossing or approaching orbits. *Mon. Not. R. Astron. Soc.* 423: 1674–1690.

Gehrels, T. 1994. *Hazards Due to Comets & Asteroids*. Tucson, AZ: University of Arizona Press.

Grieve, R.A.F., Dence, M.R. 1979. The terrestrial cratering record II. The crater production rate. *Icarus* 38: 230–242.

Hahn, G., Bailey, M.E. 1990. Rapid dynamical evolution of giant comet Chiron. *Nature* 348: 132–136.

Hills, J.G., Nemchinov, I.V., Popov, S.P., Teterev, A.V. 1994. Tsunami generated by small asteroid impacts. In *Hazards Due to Comets & Asteroids*, ed. T. Gehrels, 779–789. Tucson, AZ: University of Arizona Press.

Holloway, N.J. 1997. Tolerability of risk from NEO impacts. *Spaceguard Meeting at Royal Greenwich Observatory*, Cambridge, July 10, 1997.

Hoyle, F. 1993. *The Origin of the Universe and the Origin of Religion*. Rhode Island: Moyer Bell.

Hoyle, F., Wickramasinghe, N.C. 1978. Comets, ice ages, and ecological catastrophes. *Astrophys. Space Sci.* 53: 523–526.

Ivanov, B.A., Neukum, G., Bottke, W.F., Hartmann, W.K. 2002. The comparison of size-frequency distributions of impact craters and asteroids and the planetary cratering rate. In *Asteroids III*, eds. W.F. Bottke et al., 89–101. Tucson, AZ: University of Arizona Press.

Jeffers, S.V., Manley, S.P., Bailey, M.E., Asher, D.J. 2001. Near-Earth object velocity distributions and consequences for the Chicxulub impactor. Rhode Island: *Mon. Not. R. Astron. Soc.* 327: 126–132.

Lewis, J.S. 1997. *Mining the Sky: Untold Riches from the Asteroids, Comets, and Planets*. Reading, MA: Helix Books, Addison-Wesley.

Marsden, B.G. 1989. The sungrazing comet group. II. *Astron. J.* 98: 2306–2321.

Morbidelli, A., Bottke, W.F., Froeschlé, C., Michel, P. 2002. Origin and evolution of near-Earth objects. In *Asteroids III*, ed. W.F. Bottke et al., 409–422. Tucson: University of Arizona Press.

Morrison, D., Chapman, C.R., Slovic, P. 1994. The impact hazard. In *Hazards Due to Comets & Asteroids*, ed. T. Gehrels, 59–91. Tucson, AZ: University of Arizona Press.

Napier, W.M. 2015. Giant comets and mass extinctions of life. *Mon. Not. R. Astron. Soc.* 448: 27–36.

Napier, W.M., Asher, D.J., Bailey, M.E., Steel, D.I. 2015. Centaurs as a hazard to civilization. *Quarterly Jl. Roy. Astron. Soc.* 56: 6.24–6.30.

Napier, W.M., Clube, S.V.M. 1979. A theory of terrestrial catastrophism. *Nature* 282: 455–459.

NASA JPL CNEOS Website 2018. NASA Jet Propulsion Laboratory Center for Near Earth Object Studies. Discovery Statistics. https://cneos.jpl.nasa.gov/stats/totals.html (accessed February 7, 2018).

Öpik, E.J. 1966. Sungrazing comets and tidal disruption. *Ir. Astron. J.* 7: 141–161.

Rabinowitz, D., Bowell, E., Shoemaker, E., Muinonen, K. 1994. The population of Earth-crossing asteroids. In *Hazards Due to Comets & Asteroids*, ed. T. Gehrels, 285–312. Tucson, AZ: University of Arizona Press.

Rabinowitz, D., Helin, E., Lawrence, K., Pravdo, S. 2000. A reduced estimate of the number of kilometer-sized near-Earth asteroids. *Nature* 403: 165–166.

Snodgrass, C., Fitzsimmons, A., Lowry, S.C., Weissman, P. 2011. The size distribution of Jupiter family comet nuclei. *Mon. Not. R. Astron. Soc.* 414: 458–469.

Steel, D.I. 1995. Collisions in the solar system—VI. Terrestrial impact probabilities for the known asteroid population. *Mon. Not. R. Astron. Soc.* 273: 1091–1096.

Steel, D.I., Asher, D.J., Clube, S.V.M. 1991. The structure and evolution of the Taurid Complex. *Mon. Not. R. Astron. Soc.* 251: 632–648.

Toon, W.B., Zahnle, K., Turco, R.P., Covey, C. 1994. Environmental perturbations caused by asteroid impacts. In *Hazards Due to Comets & Asteroids*, ed. T. Gehrels, 791–826. Tucson, AZ: University of Arizona Press.

Toulmin, S.E. 1982. *The Return to Cosmology.* University of California Press.

Trujillo, C., Jewitt, D.C., Luu, J.X. 2001. Properties of the trans-Neptunian belt: Statistics from the Canada-France-Hawaii telescope survey. *Astrophys. J.* 122: 457–473.

Whipple, F.L. 1973. Note on the number and origin of Apollo asteroids. *Moon* 8: 340–345.

Yeomans, D.K., Chamberlin, A.B. 2013. Comparing the Earth impact flux from comets and near-Earth asteroids. *Acta Astronaut.* 90: 3–5.

CHAPTER 3

Space Weather

The Sun as a Natural Hazard

Mike Hapgood

CONTENTS

OVER THE PAST FEW decades we have become much more aware that space is a source of hazards which can adversely affect human activities. Some of these arise from our activities in space as described in other chapters of this book. Others arise from natural phenomena in space and their adverse effects on the environments in which we live and work. Space weather falls firmly in this latter category. It is concerned with a number of natural environments that are influenced by phenomena originating in space and which can have adverse impacts on many of the technologies underpinning everyday life on Earth.

3.1 SPACE WEATHER AND THE RADIO SPECTRUM

One simple example is the natural electromagnetic environment of radio frequencies. Modern societies rely on a huge range of radio technologies for communication and surveillance, working over frequencies from a few tens of kilohertz up to many tens of gigahertz. The success of these technologies relies, in turn, on the lack of interference from natural radio emissions across these frequencies. But there are sometimes natural radio sources which do cause interference. Some come from phenomena on Earth, such as lightning, but some come from space. One example is the aurora, whose visible light is produced by energetic electrons bombarding the upper atmosphere at altitudes of 100 to 200 km. At higher altitudes (6000 to 10,000 km), those same electrons generate radio emissions at frequencies of 50 to 500 kHz (known as auroral kilometric radiation or AKR). These are very strong emissions with a total power of several gigawatts, which can be observed at great distances from the Earth (Lockwood 2007). Fortunately, these emissions cannot reach the surface of the Earth, as the ionosphere—the ionized part of Earth's upper atmosphere—blocks radio waves at frequencies below a few MHz; an effect which allowed the development of the first radio broadcast services from the 1920s to the 1960s, operating at frequencies between 200 and 1000 kHz. Indeed, AKR was only discovered in 1974 when observations were first made above the ionosphere (Gurnett 1974).

The Sun is also an important source of strong radio emissions. These are often seen by ground-based radio telescopes over frequencies from a few MHz up to several GHz. Their ability to interfere with radio systems was first recognized during World War II through impacts on radar (Hey 1946) and continues to this day. A strong solar radio burst on November 4, 2015 was identified as the key cause of radar problems which led to a 2-hour shutdown of airspace over the whole of southern Sweden, causing a huge disruption of air traffic over Sweden and many surrounding countries. Solar radio bursts are a particular important threat to satellite navigation and timing systems, such as GPS, as these systems necessarily rely on weak signals (a 500 W transmitter on the satellite gives about 10^{-15} W cm^{-2} on the surface of the Earth). Thus, strong solar radio bursts can jam the reception of GPS signals across the dayside of the Earth. Fortunately, this source of jamming lasts only a few minutes, after which normal service resumes. However, this therefore highlights that all users of satellite navigation and timing services must be able to cope with interruptions of at least a few minutes through the use of complementary techniques for navigation and timing.

3.2 THE ATMOSPHERIC RADIATION ENVIRONMENT

Space weather also has a major influence on the natural radiation environment in Earth's atmosphere and on the surface of the Earth. In the normal course of everyday life human beings are exposed to low levels of radiation, mostly arising from radon gas seeping out of the ground (Public Health England 2014), but about 8% comes from cosmic rays: protons and heavier ions accelerated to very high energies (>1 GeV) by shock waves from supernovae elsewhere in our Galaxy. These particles pervade interstellar space, trapped for thousands of years in the Galactic magnetic field before being lost from our Galaxy. Most cosmic rays entering the Solar System are scattered back into interstellar space by turbulence in the solar wind—the tenuous plasma which flows out of the Sun and that fills all interplanetary space. But some cosmic rays (perhaps 25%, but variable over the solar cycle) reach the inner Solar System and enter Earth's atmosphere where they collide with atmospheric species producing the range of particles, such as neutrons and muons, which we observe on the ground and on an aircraft as "cosmic rays." While this provides only 8% of the radiation at the Earth's surface, its contribution rises rapidly with altitude as the shielding provided by Earth's atmosphere decreases. Cosmic rays are the dominant source of radiation above 3 km altitude and can lead to significant radiation doses at aircraft flight altitudes. For this reason, aircrews are recognized as being occupationally exposed to radiation (ICRP 1991) alongside workers in the nuclear industry, underground facilities, and medical diagnostics. Aircrew cosmic ray doses vary with solar activity (highest doses when activity is low, leading to least turbulence in the solar wind) and with the latitude of their flight path, as the geomagnetic field provides a degree of radiation shielding (highest doses at high latitudes). For example, during the last solar minimum in 2008–2009, travel on air routes reaching high latitudes, such as London to Los Angeles, exposed aircrews to doses of up 50 to 100 microsieverts per one-way journey; even higher doses may be expected during the next solar minimum around 2020 due to the general decline of solar activity (Miyake et al. 2017).

Cosmic rays are occasionally enhanced by solar energetic particle (SEP) events—intense bursts of protons and ions lasting from a few hours to a few weeks (Figure 3.1). When these bursts contain significant fluxes above 400 MeV they can enhance the neutron and muon fluxes reaching Earth's surface. Over seventy of these "ground level enhancements" have been observed since 1942, and in the worst case produced a 50-fold enhancement in the energetic neutron fluxes observed on the ground (Gold and Palmer 1956; Marsden et al. 1956). It is estimated that the corresponding radiation levels at modern aircraft flight altitudes were 300 times greater than normal (Dyer et al. 2007). A similar event today would be a significant concern as it could deliver radiation doses of up to 2000 microsieverts for aircrews and passengers. While these doses would be far below the 1 sievert level at which a short-term radiation dose poses an immediate threat to human

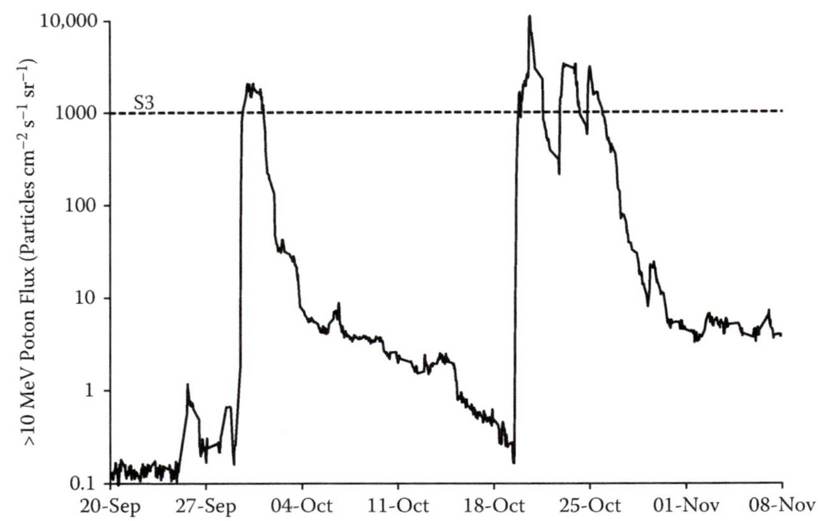

FIGURE 3.1 Major radiation storms of autumn 1989, as shown by the flux of protons at energies >10 MeV, as derived from the GME instrument on NASA's IMP-8 satellite. Two major events start on September 29 and October 19, respectively. (Figure by the author; data source: NASA CDAWeb at https://cdaweb.sci.gsfc.nasa.gov, dataset I8_H0_GME.)

health (Public Health England 2011), they would be significant for long-term health. Thus, the medical response to aviation radiation exposure during a space weather event is expected to include an assessment of the exposed individuals: First what dose did they receive? And then should they take action to manage their future radiation exposure? The underlying principle is to keep their lifetime radiation exposure as low as reasonably achievable (ALARA). This ALARA principle is central to managing the long-term risk of cancer (CDC 2015).

The natural radiation environment can also disrupt electronic systems through single event effects (SEEs)—adverse effects which can arise when a single energetic particle (a neutron or an ion) passes through an electronic device such as computer memory or a data processing unit. The particle will create a trail of ionization within the device, knocking electrons out of their orbits around the nuclei of the atoms. They can disrupt the device if there are sufficient numbers of these free electrons. At its simplest this disruption can change a computer bit—the zeroes and ones which store data and codes in digital devices. However, these "single event upsets" are straightforward to correct either by use of error correction codes, which can detect and fix such changes. Single event upsets only become a serious problem when a large numbers of events occur around the same time, overloading the capabilities of correction codes.

But single event upsets are just one of a menagerie of disruptive effects, which has grown as the sizes of features within electronic devices have been reduced over the past 50 years. Single event effects now include: (1) transients, where ionization trails generate strong electric fields misinterpreted as signals; (2) latch-up, where devices become locked into a particular state; and (3) gate rupture, where the ionization trail creates a conducting path through a semiconductor gate resulting in catastrophic failure of that element of the device (Pisacane 2016). There is significant concern that severe space weather in the form of an intense atmospheric radiation event could lead to high SEE rates in the electronic systems which now control aircrafts; therefore, those systems need to be resilient to SEEs. This was well demonstrated in 2008 when a control system malfunction caused abrupt changes in the attitude of an aircraft, causing serious injuries to 11 passengers and one crewmember. The precise cause of the malfunction was never established but the official report (ATSB 2011) reinforced the importance of ensuring that aircraft systems deal robustly with such malfunctions, including SEEs. This should be verified by testing in simulated radiation environments (Hambling 2014; STFC 2017).

Single event effects are also a growing concern for electronic systems used in many human activities on the surface of the Earth. The existence of such effects was demonstrated during a 2003 election in the Belgian town of Schaerbeek, where a voting machine recorded an extra 4096 votes for one candidate, more than the number of votes that could have been cast in the election. The error was attributed to a single event upset flipping the 13th bit in the counter (PourEVA 2003). Another example was a series of incidents in which electronic power control devices failed despite being relatively new (Normand 1997). This was traced to single event effects triggering burnout of the devices and was resolved by redesign of the devices. There are probably many other adverse impacts on electronic devices which go unreported and that are attributed to unknown malfunctions. Obviously, a fraction of the cases of computer malfunction is likely due to single event effects from atmospheric radiation, and often these problems are simply cleared by rebooting the computer. There is statistical evidence that such problems are more common in areas where people live and work at altitudes well above sea level (Normand 1996) where they are more exposed to radiation from space.

However, the principal concern about single event effects in ground-based systems is their potential to disrupt electronic devices which control systems directly critical to human safety or that are societally critical infrastructures (electricity, water, gas, landline and mobile communications, rail, and road transport). The big question is whether a severe space weather event such as observed in 1956 could produce high rates of single event effects leading to disruption of these systems. This area is gradually gaining attention, for example through studies on the impact of a space weather event on the control systems for nuclear power stations. The operators of these systems have a requirement to consider extreme events at the 1 in 10,000 year level (HSE 1992) and hence need to consider the impact of extreme radiation storms. This includes not just those of the severity of the 1956 event discussed above, but also more extreme events such as storms thought to have occurred in AD 774/775 (Miyake et al. 2012). This event was first identified from a high concentration band of C^{14} deposited in dated tree rings in Japan, which was later detected at many other sites around the world. Analysis of the event suggests that it was produced by an intense radiation storm, which included high fluxes of very energetic (GeV) particles (Thomas et al. 2013) capable of reaching the ground as well as the altitude between 10 and 15 km where C^{14} is naturally produced by space radiation. An intense space weather event on the Sun is considered the most likely source of the event (Usoskin et al. 2013). Other space weather events such as a nearby supernova have been proposed but lack crucial evidence, for example, no suitable supernova remnant exists in our extensive observations of nearby remnants.

These studies of single event effects need to be extended to cover many systems outside the nuclear industry. For example over the past 20 to 30 years, many critical infrastructures have implemented electronic control systems based on an architecture called SCADA (Supervisory Control and Data Acquisition). This has enabled distributed infrastructures to be operated remotely from a few control centers and has reduced the need to position staff across the infrastructures as part of the control system. However, we do not understand today how SCADA-based control systems would respond to a severe space weather event, let alone how to prepare for such an event. It is possible, indeed likely, that such an event would be a temporary disruption lasting an hour or two, and that the problem would go away once the affected devices are reset. But without analysis and preparation such an event could be dangerous as people would respond without understanding. In contrast, understanding would allow people to wait in safe places and let the storm pass, while allowing operators to be prepared to reset devices, for example, through provision of complementary systems to enable resets.

3.3 SPACE WEATHER AND SATELLITES

3.3.1 A Range of Space Environments

Space weather, quite naturally, has a significant influence on the environment in which satellites operate. This leads to a number of significant effects—as we shall discuss below—but sometimes also to a mistaken belief that the most important space weather effects are on satellites since they are more directly exposed to it. In practice, the most dangerous space weather effects arise on Earth when the magnetosphere and ionosphere focus energy flowing from the Sun to produce geomagnetic storms and substorms, leading to major disruptive effects on power grids, communications, and transport systems. Satellites are less exposed to this focusing and benefit from good awareness of the problem among their designers and are built to withstand the rigors of the environment in space, including the effects of space weather. Major space agencies such as ESA and NASA have long supported "space environment and effects" groups as a focus for expertise on space weather impacts on satellites. As a result satellites are explicitly designed to have a high resilience to space weather.

3.3.2 Radiation Effects on Satellites

Radiation is the most obvious space weather effect on satellites. Cosmic rays and energetic particles from the Sun impact satellites, just as they impact the Earth, and produce single event effects in electronic devices. Satellite designers mitigate these effects by good engineering, for example, radiation shielding around vital electronics, error correction codes to clear single event upsets, and parallel redundancy to work around disruptions of any particular device. But this mitigation is never perfect. Single event effects are a significant cause of satellite anomalies, and hence require action from satellite operators. Even in quiet space weather conditions they are a substantial part of the workload of satellite operators and become a major burden during severe radiation storms, such as those in the autumn of 1989 shown in Figure 3.1. During these storms there were considerable periods when satellites were likely to experience single event effects due to particle radiation fluxes above the S3 level (NOAA 2011), as marked by the horizontal dashed line in Figure 3.1. The first storm in late September exceeded S3 for a single continuous period of 38 hours, while the second storm in mid-October exceeded S3 for 67 out of 140 hours. A similar event today would inevitably pose a challenge as operations teams work around the clock to maintain satellite services. Fortunately, these events have helped to drive the modern design standards for satellite resilience against radiation and are a vital counterbalance to current operational experience, which will inevitably focus on the relatively benign radiation conditions we have experienced over the past decade.

Radiation damage is also important for satellites. This is a nonionizing effect, which arises as energetic ions penetrate into materials. Coulomb interactions between these energetic ions and the nuclei of atoms in satellite materials can displace some of those nuclei from their usual positions. The growth of such displacements will gradually degrade the performance of devices such as solar arrays and electronic chips. Over many years, radiation damage can accumulate to a point where devices fail. Satellite designers will again mitigate these effects by good engineering: radiation shielding and the choice of less susceptible materials. This mitigation can extend satellite lifetimes, delivering more value for the satellite operator whether in industry, science, or government. More severe space weather will likely result in rapid aging of satellites. Thus, end-of-life planning for satellites should take into account the possibility of older satellites being brought to their end by radiation damage during severe space weather.

3.3.3 Charging Effects on Satellites

Space is never a true vacuum. It is pervaded by plasma, matter in which neutral atoms have been broken down (ionized) into electrons and ions. The plasma may be so tenuous that it is less dense than what passes for vacuum in a terrestrial laboratory. But the plasma is there nonetheless and will interact electrodynamically with satellites, so that they become electrically charged. This interaction with the plasma takes two distinct forms: (1) one with the bulk plasma, leading to charging of satellite surfaces; and (2) one with the high-energy electron population of the plasma, leading to charging inside the satellite.

3.3.3.1 Surface Charging

Surface charging arises on all objects in space as they exchange electric charge with the local plasma environment, in particular due to photoelectron emission when the surface is exposed to solar ultraviolet radiation, and due to bombardment of the surface by the plasma particles. Electrons generally dominate this bombardment because their mass (m_e) is much less than that of ions. For a simple plasma of protons and electrons with both particle distributions having the same mean energy (temperature), the mean velocity of the electrons is higher than that of the protons by a factor $\sqrt{m_p/m_e} \approx 43$, where m_p is the proton mass. Since the number density of electrons and protons must be the same (to ensure electrical neutrality of the plasma), this implies that the plasma electron fluxes bombarding the surface of the object are 43 times greater than the proton fluxes. Hence, the plasma bombardment deposits electrons on to the surface while photoemission removes electrons from the surface.

The surface of the object will acquire an electric potential relative to the local plasma such that the net flow of electrons in and out of the surface is zero (Whipple 1981). In sunlight the photoemission tends to dominate, so that the surface acquires a positive potential sufficient to prevent photoelectrons escaping from the surface. This is typically a few volts positive and corresponds to the typical photoelectron energies of a few electron volts. But if the electron bombardment dominates (most obviously when the satellite is in darkness) the surface will acquire a negative potential sufficient to repel plasma electrons with energies close to the mean of the energy distribution. In some cases this can require a negative potential of thousands of volts corresponding to plasma mean energies (temperatures) of thousands of electron volts.

These potentials do not pose a dangerous risk if they are distributed evenly across the whole body of the satellite (a condition known as frame or overall charging), as should occur if there is adequate electrical connectivity across the surfaces and body of the satellite. The danger comes if there is differential charging of surfaces or other elements of the satellite, for example, on exposed dielectric materials that cannot release their charge to the frame of the satellite, leading to potential differences of thousands of volts between different elements of the satellite.

Surface charging is a significant risk for satellites in geostationary orbit. This orbit lies in the outer part of Earth's magnetosphere, which is pervaded by plasma with mean energies of thousands of electron volts. Charging is also a significant risk for satellites in low (500 to 1000 km altitude) orbits crossing polar regions, where they will be bombarded by the auroral electrons with energies of thousands of electron volts.

In principle, ensuring that all satellite surfaces are electrically interconnected can mitigate surface charging. Modern satellite builders make major efforts to achieve this by adding conducting layers and paths. For example, multilayer insulation (the gold foil wrapped around many satellites) has a conducting surface to ensure this, and is electrically connected to the satellite frame. In addition, where surfaces must be made from dielectric materials, it is desirable to apply a thin conductive coating and provide a conductive route from the coating to the satellite frame. However, the mitigation of surface charging will always require a trade-off against some of the other requirements on the design of satellites. For example conductive coatings on dielectrics, such as lenses, may reduce their performance and consequently there will have to be engineering compromises, which balance the performance against the charging risk.

3.3.3.2 Internal (Deep Dielectric) Charging

Internal charging of satellites arises when particles have sufficiently energy to penetrate and deposit electric charges deep inside satellites, for example, by high energy (MeV) electrons in the Earth's magnetosphere. These electrons are dangerous when they deposit a charge deep inside dielectric materials such as circuit boards and electrical insulation. In these cases the charge will leak away only very slowly and—if accumulated over several days—it can produce strong electric fields, which can cause an electrical breakdown inside the dielectric (as demonstrated in benchtop experiments on Earth; a wonderful visual example is available online at https://youtu.be/

eCz7BL74D4Y). These discharges can generate false signals which disrupt satellite operations and even directly damage satellite systems.

This deep dielectric charging is difficult to mitigate by design as it is not straightforward to provide conductive paths that allow charge to leak away, though shielding around vulnerable dielectrics can slow charging rates. Thus, the occurrence of high fluxes of MeV electrons is a concern to satellite operators. These "killer electrons" follow a distinctive temporal variation in which there is a rapid rise in fluxes over a range of distances from Earth and then a slow decay. They are a serious threat to satellites in geostationary orbit at altitudes of 36,000 km and in "middle Earth orbit" at 20,000 km altitude used by global navigation satellites such as GPS and Galileo.

Thus, it is helpful to understand and forecast when these high fluxes will occur; there are two key factors to consider:

1. The speed of the solar wind. High solar wind speeds (>600 km s^{-1}) impinging on the Earth generally lead to high electron fluxes (Paulikas and Blake 1979). The turbulent flow of high-speed solar wind past Earth is thought to generate low frequency (kHz) plasma waves in the magnetosphere, which can accelerate electrons to very high energies (Horne et al. 2005).

2. The natural variation of high electron fluxes according to their position within the Earth's magnetic field, as specified by a parameter called L-value (McIlwain 1961). This parameter has a sophisticated physical definition, but for our purposes here it can be interpreted as a measure of the geocentric distance, measured in Earth radii, at which the trajectory of an energetic particle trapped in Earth's magnetic field crosses the magnetic equator. Energetic electron fluxes tend to peak around an L-value of 4 (Baker et al. 1994) close to where satellites in middle Earth orbit cross the magnetic equator, so these experience the highest fluxes. Satellites in geostationary orbit at around L = 6.6 will see lower but still significant fluxes. Furthermore, since geostationary orbit is aligned with the geographic equator, the L value around that orbit will vary with longitude, exhibiting minima where it crosses the geomagnetic equator as shown in Figure 3.2. Thus fluxes, and the charging risks, are highest in those regions around minimum L value regions, around 20°E and 160°W (Meredith et al. 2015).

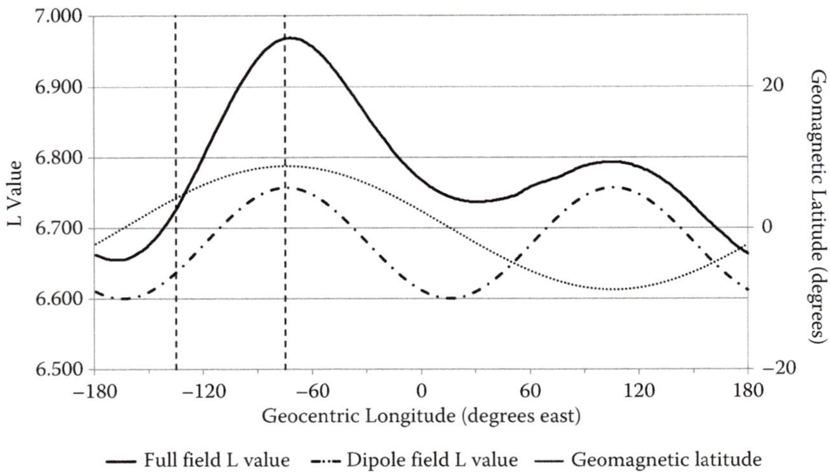

FIGURE 3.2 Variation of L-value (heavy solid curve) and geomagnetic latitude (dotted curve) with geocentric longitude around geostationary orbit. The two vertical dashed lines at 135° west and 75° west show the approximate locations of NOAA's GOES West and GOES East satellites. The asymmetry in the L-value curve arises mainly because the Earth's magnetic dipole is displaced from the center of the Earth toward East Asia. For comparison the dash-dot curve shows how L-values would vary in a centered dipole field. (Figure by the author.)

3.3.4 Satellite Drag

All planetary atmospheres exhibit decreasing density with increasing altitude so that the atmosphere gradually thins as one travels out into the space environment around the planet. However, this thinning atmosphere is sufficiently dense to create a small drag force on a satellite as it ploughs its way through that atmosphere. This drag force causes the satellite to move closer to the planet, losing some of its gravitational potential energy and decreasing its orbital period. This process will eventually bring the satellite to an altitude where the drag is so large as to cause immediate reentry.

In the case of the Earth, satellite drag is dependent on the density of matter in the thermosphere, the neutral part of Earth's upper atmosphere at altitudes between 200 and 800 km. This, in turn, is dependent on the temperature of the thermosphere because the temperature controls the rate at which density decreases with altitude. When the temperature is high, the rate of decrease is smaller and hence one has higher densities (and greater drag) at higher altitudes. In quiet space weather conditions, the temperature is controlled by the extreme ultraviolet (EUV) irradiance from the Sun, with highest temperatures when irradiance is high, such as at solar maximum, and lowest temperatures when irradiance is low, such as at solar minimum. This has the important consequence that satellite drag is greater at solar maximum and hence more old satellites will reenter the atmosphere during solar maxima. This is an important feature, which satellite operators can exploit in their plans for the removal of old satellites. Such removal is now an important feature in managing satellite collision risks (see Chapter 5).

Space weather has profound effects on the density of the thermosphere. In particular geomagnetic storms can cause intense heating in the auroral zone, leading to an upwelling of matter into the thermosphere and to an increase in thermospheric wind flow away from the poles, thus transporting this denser matter to lower latitudes. As a result, geomagnetic storms can significantly change the global density of the thermosphere and change the drag force on satellites.

This can significantly change satellite orbits. The initial impact of this increased density is to reduce the orbital period of the satellite by more than was expected for quiet space weather conditions. Thus, the satellite will not be located where it was expected. For example, if a satellite was predicted to pass over a particular site on Earth at time T, increased drag will cause it to pass that site at an earlier time (due to the reduced orbit period) and at some distance to the east of the site (since the Earth will not have rotated quite so far). Furthermore, this error will accumulate with time; if the timing error is Δt after orbit, it will be $N\Delta t$ after N orbits. Similarly, the satellite ground track will continue to move further east than originally expected. These changes in the satellite orbit will pose a variety of problems for satellite operators, including:

- Scheduling of operations that depend on satellite location relative to Earth, for example, contacts with ground stations, and observations of selected regions of the Earth's surface.
- Assessing the risk of collisions with other satellites and with debris from older satellites (see, e.g., Chapter 5).
- Monitoring the locations of satellites. Satellites in low Earth orbit are usually tracked with radars. But this tracking is done episodically as the satellite pass over radar sites, thus the radar operators will try to correlate each radar track with a known object. After a severe space weather event, this correlation is much more difficult due to major changes in the orbits, and it can take several days for many satellites

to be matched with their changed orbits. For example, following the great geomagnetic storm on March 13/14, 1989, the U.S. Air Force lost track of over 1500 satellites (Air University 2003).

- Monitoring satellites close to reentry. The increased satellite drag arising from a severe space weather event will complicate the management of risks rising from reentry of large satellites; it could cause reentry to occur early and in an unexpected location.

3.4 THE IONOSPHERE—AT THE HEART OF SPACE WEATHER IMPACTS

3.4.1 How Is the Ionosphere Formed

Ionization in the upper atmosphere arises primarily from the action of solar EUV (10–100 nm) on the neutral gas in the upper atmosphere producing a layer of plasma, the F-region, extending from 150 km to 500 km altitude. At these altitudes the conversion of plasma back to neutral gas ("recombination" of ions and electrons) is usually a slow process, thus F-region plasma produced in the daytime will usually persist throughout the subsequent night. This is a key feature of the ionosphere, one of wide importance for the operation of radio technologies. For example, it is exploited in systems that use MHz radio signals for long-distance radio communications; at these frequencies the ionosphere will reflect radio waves, so MHz radio can propagate from one location on Earth to another far beyond the horizon. This technology dates back to the 1920s, but is still widely used in civil aviation and by the military. The nighttime persistence of the F-region allows it to operate at all times of day.

As a result of the ionization which produces the F-region, very little EUV penetrates below 150 km altitude. However, solar soft X-ray emissions (1 to 10 nm) can penetrate to lower altitudes and produce the E-region of the ionosphere, extending down from 150 to 90 km altitudes. At these altitudes, the plasma is rapidly (tens of seconds) destroyed by recombination. Thus, the E-region is a daytime phenomenon, appearing soon after sunrise and disappearing soon after sunset. It is therefore of lesser importance for practical purposes.

The Sun occasionally produces strong fluxes of hard X-rays (0.01 to 1 nm), which can penetrate below 90 km altitude producing a further layer of ionization termed the D-region. At these altitudes the free electrons in the plasma experience frequent collisions with atmospheric species; as a result radio waves can be absorbed by the plasma, causing a blackout of radio communications at MHz frequencies. Thus, these communications are at high risk of interruption when the Sun produces intense bursts of very short wavelength X-rays in the form of solar flares.

The D-region ionosphere can also be enhanced when solar energetic particles are guided by geomagnetic fields into the polar atmosphere and create ionization through collisions with atmospheric species. Thus, SEP events can lead to a blackout of MHz radio communications over polar regions, an effect known as a polar cap absorption (PCA) event. These events have major practical impacts on trans-Arctic air routes which link major population centers in Asia and North America. Since continuous radio contact with ground control centers is a mandatory requirement for civil aviation, and current satellite communications systems do not work north of 82° latitude, these routes must be closed when a PCA blacks out MHz radio communications. The cost of such a disruption has been estimated at $100,000 per diverted flight (Space Studies Board 2009).

3.4.2 An Outline of Ionospheric Morphology

The overall morphology of the ionosphere is determined only in part by the processes that create ionization. Two other factors are important, most obvious are the recombination processes which destroy ionization, but transport also plays an important role since the majority of ionospheric plasma is located in the F-region, where recombination is usually slow, therefore the ionospheric plasma has a lifetime of many hours and can be transported hundreds or even thousands of kilometers. This transport leads to the formation of regions of both enhanced and depleted plasma density in different parts of the world, as transport moves plasma into regions where loss rates are lower or higher.

One example of this transport occurs in the polar ionosphere, where regions of dense and relatively cool plasma are observed to drift from the dayside to the nightside. These "polar patches" typically have horizontal scales of a few hundred kilometers. They arise because the flow of ionospheric plasma across the polar region is driven by momentum from the solar wind (Axford and Hines 1961). But this flow is episodic as magnetopause reconnection leads to bursts of solar wind momentum being transported down magnetic field lines to the polar ionosphere. Each burst will draw dense F-region plasma away from the dayside producing a separate patch, which then drifts over the nightside. Polar patches have significant impacts on radio systems operating in polar regions: they change the radio frequencies used in aircraft communications and also the ionospheric corrections required by radar and global navigation satellite systems (GNSS) operating in polar regions. Another important example of transport affecting ionospheric morphology comes in equatorial regions, more precisely in two regions north and south of the magnetic equator where the magnetic field direction has a gentle slope relative to the horizontal (at the equator it is horizontal). Momentum transfer from neutral winds to ions can drive significant plasma flows, which must then follow the geomagnetic field direction, leading to an uplift of plasma as winds flow toward the equator. This will increase plasma lifetime (as the recombination rate decreases with altitude) and hence increase plasma density. The net result is to generate a region of enhanced plasma density either side of the magnetic equator: the equatorial ionization anomaly.

3.4.3 Plasma Instabilities and Ionospheric Scintillation

Plasma instabilities can cause the ionosphere to break down into a series of filaments, some with high density and some with low density, which will align along the local magnetic field direction. As a result the ionosphere will act as diffraction screen for radio waves, thus radio signals passing through the ionosphere show variations in amplitude and phase, an effect known as scintillation. This is exactly analogous to the twinkling of stars in the night sky, where turbulence in the lower atmosphere acts as a diffraction screen for the light from the stars. Ionospheric scintillation of radio signals is a significant problem for radio frequencies from a few MHz up to about 3 GHz. In particular, it can disrupt radio links to and from satellites including important applications such as satellite navigation (e.g., GPS and Galileo) and many satellite communications systems.

There are two main regions in which ionospheric scintillation occurs. One is the auroral zone where ionospheric density exhibits strong spatial gradients which can be a source of plasma instabilities; the other is equatorial regions around dusk where we find upward plasma flows which can breakdown via the Rayleigh-Taylor instability.

3.4.4 Storms in the Ionosphere

Geomagnetic storms can lead to major changes in the overall structure of the ionosphere. In the early phases of a geomagnetic storm, the electric field in the solar wind will penetrate the magnetosphere and reach down to the ionosphere on the dayside of the Earth. This electric field must be eastward as it arises from the southward interplanetary magnetic field needed to drive a geomagnetic storm. This will drive an uplift of ionospheric plasma on the dayside of the Earth, leading to enhanced plasma densities in the ionosphere. As the geomagnetic storm proceeds, the systems of electric currents in the magnetosphere will intensify and exclude the solar wind electric field—hence this first enhancement of plasma densities in the dayside ionosphere will subside. But it is likely to be followed some hours later by a further enhancement as auroral heating drives thermospheric winds toward the equator, and hence upward flows of plasma along geomagnetic field lines. These enhanced densities in the early phase of the storm form the positive phase of the ionospheric manifestation of the storm. But this is then followed by a negative phase in which plasma densities are heavily depleted as auroral heating drives an upflow of molecular species (such as nitric oxide, NO) to altitudes above 150 km, after which thermospheric winds will transport them across the globe. This will increase the loss rates of ionization in the F-region leading to a rapid decay of its plasma density after sunset. This is a characteristic feature of the negative phase of an ionospheric storm—the F-region largely disappears during the first night after the storm starts. It will reappear the next day, but usually with lower densities than normal and then disappear again the next night. This cycle of no nighttime F-region and weak daytime F-region may continue for some days, showing a gradual recovery toward normal conditions. This reflects a gradual return to normal thermospheric composition and loss rates, and demonstrates how ionospheric storm effects can persist after the geomagnetic storm. An example of these effects is shown in Figure 3.3.

These changes in F-region plasma density, especially the nighttime depletion, have profound impacts on many radio systems. For example, MHz radio communications will become impossible when the F-region disappears at night. This disappearance may also lead to errors in satellite navigation systems (GNSS). To determine their position on Earth, these systems first measure the travel time of radio signals between several GNSS satellites and the receiver location; then, knowing the signal speed, they can calculate the distance between each satellite and that location, and finally determine the position on Earth that is most consistent with these distances. The critical issue is what is the signal speed; the radio signals pass through the ionosphere where they are subject to a slight delay (compared to travel at the speed of light) because the plasma in the ionosphere has a group refractive index slightly great than one. This delay is proportional to the total electron content (TEC—the column density of plasma along the signal path) and can lead to distance errors of tens of meters. Thus, if the TEC is known one can significantly improve the accuracy of these position

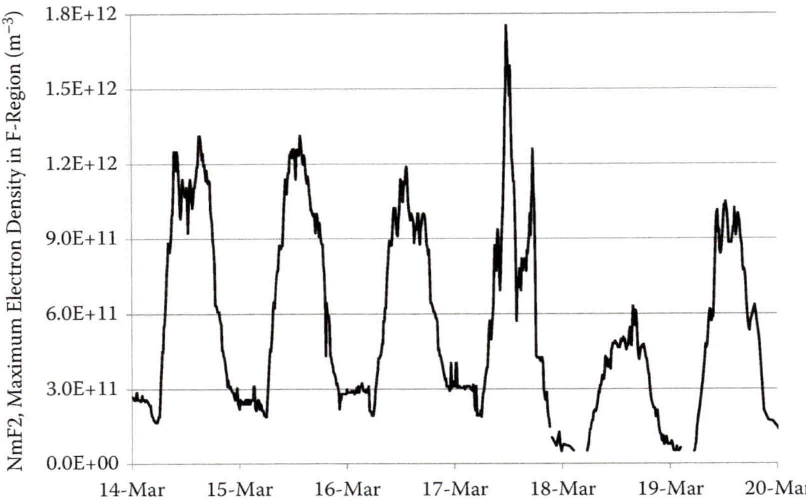

FIGURE 3.3 Density changes in the ionosphere over southern England on March 14–19, 2015. The first three changes show a typical diurnal pattern with density rising at dawn and falling at dusk, but significant density persisting through the night. On March 17, a major geomagnetic storm starting at 04:45 UTC gives two distinct daytime enhancements of plasma density, followed by very low densities during the subsequent night. Daytime densities are much more reduced on March 18, and followed by another night of low densities. (Figure by the author; densities derived from Chilton ionosonde data obtained from the UK Solar System Data Centre.)

determinations. This is very important for some navigational applications, for example, in civil aviation, which has led governments around the world to invest billions of dollars in the provision of systems to determine and disseminate values of these ionospheric corrections and to distribute them to satnav users (Loh et al. 1995; Gauthier et al. 2001; Rao 2007; Trinity House 2016; Avanti 2017).

Modern correction systems such as the European Geostationary Navigation Overlay Service (EGNOS) (Gauthier et al. 2001) can detect and warn of major changes of TEC on timescales less than 10 seconds. Thus, these systems are able to adapt for space weather changes in TEC during most storms. However, in the most severe storms, TEC will vary rapidly in both space and time such that the correction systems cannot keep track of the changes. In these cases, the correction systems should detect this problem and warns users not to rely on the correction.

These correction systems can deal very well with the changes in TEC caused by solar flares. In these cases TEC exhibits a very smooth spatial variation across the Earth and rapid temporal changes only at flare onset. As a result, the effect of solar flares on satellite navigation is almost completely mitigated by systems such as EGNOS.

Another important element of a major storm is that the aurora will move down to lower latitudes. For example, it extended down to the south of England and northern France during the great geomagnetic storm of March 1989. Unfortunately, this expansion does not just bring the beautiful display of the aurora to these lower latitudes, it also brings some of adverse space weather impacts which are usually confined to polar regions. In particular, it will bring the high latitude scintillation which can disrupt satellite navigation and communications services (Cannon et al. 2013). During a severe storm it is likely that these services will be disrupted for several days.

In summary, ionospheric storms are a major manifestation of space weather and have major impacts on a wide variety of radio systems. In particular, they can cause major disruption of MHz communications (used by civil aviation and the military), of satellite communications (used by a wide community including aviation and shipping), of satellite navigation (used by transport services across the world), and even the VHF satellite systems used to collect tracking data from ships out of sight of land. Thus, ionospheric storms pose a major economic and societal risk, one that is perhaps underappreciated today because the recent growth in the use of many of the systems has been accompanied by a decade without major storms (as shown in Figure 3.4).

3.5 GEOELECTRIC FIELDS—THE THREAT TO GROUNDED INFRASTRUCTURES

Space weather induces electric fields in the solid body of the Earth and thus can drive electric currents through any infrastructure electrically connected to the Earth. These connections provide electrical grounding to ensure that zero electric potential in the infrastructure matches the electric potential of the Earth's surface near it. This alignment of zero potential is fundamental to electric safety as it ensures that exposed metal surfaces do not pose any risk to people. It also helps to minimize the electrical interference generated by the infrastructure, enabling it to be a good neighbor to other electrical systems.

Thus, these ground connections play a key role in the resilient operation of many infrastructures. However, where these infrastructures are spatially extended (e.g., systems such the power grid, pipelines, and railways), they

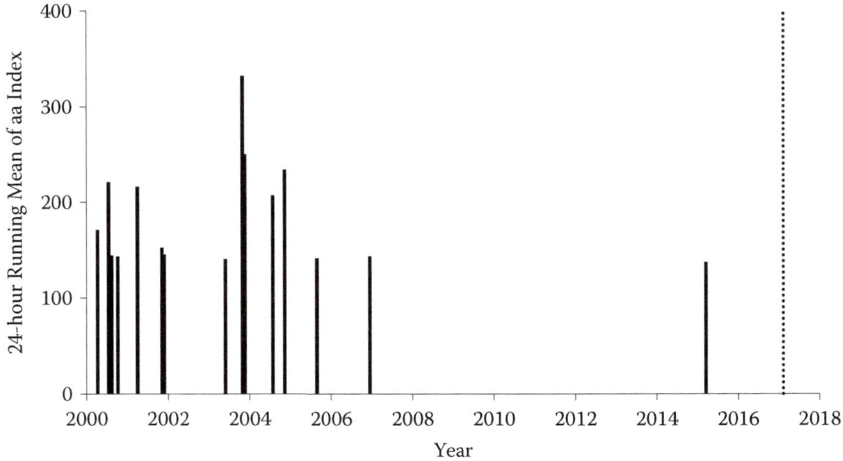

FIGURE 3.4 Absence of major geomagnetic activity since 2006, as shown by the time and strength of geomagnetic storms whose intensity matches or exceeds the St. Patrick's Day storm of 2015. The figure shows 16 storms during the period 2000–2006, some much larger than the 2015 storm. This highlights the weakness of recent geomagnetic activity compared to 2000–2006. A wider survey back to 1868 shows no comparable period of weak activity. (Figure by the author; data from UK Solar System Data Centre.)

must have multiple grounding points to ensure that the local zero potential of the infrastructure tracks any changes in ground potential. However, those multiple grounding points provide pathways by which geoelectric fields in the body of the Earth can drive electric currents through the infrastructure.

Space weather is a major generator of geoelectric fields. Time-varying electric currents in Earth's ionosphere and magnetosphere drive geomagnetic field variations, which we can observe at the surface of the Earth. These variations include significant power at low frequencies (tens of millihertz), which can penetrate hundreds of kilometers into the body of the Earth where they can induce electric fields through the action of Faraday's Law. The resulting telluric or Earth currents have been known to (and measured by) science since the nineteenth century (e.g., see Aplin and Harrison 2014). The geoelectric fields driven by space weather vary with the conductivity of subsurface rocks, such that highly resistive rocks lead to large fields while more conductive rocks (e.g., sedimentary rocks containing large amounts of water) lead to lesser fields.

By far the most important infrastructures at risk from these geoelectric fields are electric power transmission grids. Over the past 60 years these have become a fundamental infrastructure of modern societies, providing energy which is essential to the smooth operation of offices, factories, and homes as well as many other critical infrastructures for energy, transport, and communications. Loss of electric power for more than a few minutes will quickly disrupt our lives. Transmission grids enable electricity to be generated where most convenient and transported over long distances to where it is used. They are networks of electrically conducting cables linking nodes where transformers (usually more than one) convert electricity between different voltages (hundreds of kilovolts for transmission over the grid and lower voltages for distribution to users). The grid is electrically grounded through an earth connection at each transformer, therefore the geoelectric fields induced by space weather will drive electric currents into the power grid through these connections. These geomagnetically induced currents (GIC) are additional to the electric currents that carry power across the grid, and have a very low frequency (millihertz) compared to the 50 or 60 Hz alternating current in a power grid. Thus, GICs provide a DC offset which can drive transformers into half-phase saturation. This can cause some of the power passing through the transformer to be dissipated as heat and vibration, and can generate harmonics of the basic grid frequency. The heat and vibration will age a transformer more quickly than normal, reducing its productive lifetime, and in the worst cases can damage the transformer, forcing it to be removed from service. The harmonics will be transmitted across the grid where they can disrupt other devices needed for smooth operation of the grid. Half-phase saturation also acts to reduce the reactive power in a grid, the power that supports the operation of the grid, in severe cases this reduction can destabilize the operation of the grid leading to voltage collapse and power outage. This is the prime risk to power grids from a severe space weather event—a grid shutdown leaving people without power. The grid itself would suffer little permanent damage, so it should be straightforward to restart using well practiced procedures (a so-called blackstart). However, this restart would take some time and the power may be out for many hours or several days following severe space weather.

A secondary risk from severe space weather is damage to multiple transformers. A well-designed grid should be resilient to loss of a single transformer but a loss of multiple transformers will degrade its capacity to deliver power. If multiple losses include all transformers at one grid node, the result will be an extended power outage in the vicinity of that node.

While space weather impacts on power grids have been reported as far back as 1940 (McNish 1940), they only came to prominence with the voltage collapse of the Hydro-Québec power grid during the great geomagnetic storm of March 1989. In this case, GIC caused a series of problems which led to the grid going from nominal operation to fully off in just 92 seconds (Bolduc 2002). The operators at that time had no awareness of the risk and were taken completely unaware. Fortunately, they were able to restore power in just 9 hours. Nonetheless, this outage caused major disruptions with costs estimated at 2 billion dollars (Hapgood and Thomson 2010). For comparison, the event caused only limited damage to the grid, with costs estimated at 13 million dollars (Bolduc 2002). The same storm later caused two transformers in the United Kingdom to be taken out of service due to alarms indicating possible damage (Erinmez et al. 2002). The 1989 storm was a wake-up call for the power industry, stimulating a huge amount to work to understand and mitigate space weather risks to power grids (Pulkkinen et al. 2017).

A series of major space weather events in the autumn of 2003 provided another important example of how space weather can disrupt power grids—in that it led to the loss of many transformers in South Africa, thereby significantly reducing the capacity of the South African grid to provide power to its users (Gaunt and Coetzee 2007). Data from sensors on several of these transformers suggested they had suffered internal heating during that series of storms (Thomson and Wild 2010). These heating events aged the transformers so that they failed soon after the storms. These events in South Africa provide two important lessons: (1) while severe space weather does not necessarily destroy transformers, it will age transformers so that they are likely to fail much sooner than expected (Gaunt 2014); and (2) that space weather can impact on power grids in countries in low latitudes.

Another technology at risk from GICs generated by space weather is modern rail systems, where a single signaling center can control a large area via remote operation of color light signals and uses track circuits to monitor train locations. These circuits apply a small voltage between the two metal rails of any segment of the railway track. The presence of a train in that segment provides a conducting path between the two rails and the resulting current provides an electrical signal which informs the control center of its presence. It can also be used to set nearby signals to red to show that the track segment is occupied. The electric circuits that control signals and monitor track circuits are thought to be vulnerable to GICs and indeed that GIC could put signals in an incorrect state (Krausmann et al. 2015). There is circumstantial evidence for this from anomalies which occurred during major storms in 1982, 1989, and 2003 (Wik et al. 2009; Eroshenko et al. 2010). Recently, Liu et al. (2016) made the first direct measurement of GIC in rail circuits. This is a challenging task because of the complexity of these circuits and their earth connections. This is especially so if the line is equipped for electric traction, as the rail circuits may well share earth connections with power lines feeding the train engine. Despite this complexity, Liu et al. (2016) were able to demonstrate the presence of small GICs in rail circuits during a number of small geomagnetic storms in 2015. Space weather impacts on rail systems are an important area for future research, given the expanding use of high-speed rail. It is important that we understand how GICs can affect the control and safety of this technology.

3.6 THE SUN AS THE ENGINE OF SPACE WEATHER

Now that we have reviewed how the adverse impacts of space weather arise in a variety of environments around the Earth, we will discuss how those environments are influenced by phenomena on the Sun. The Sun is the primary energy source that drives space weather on Earth via a range of emissions including hot plasma, particle radiation, and electromagnetic radiation. Thus, we seek to understand how the Sun generates these emissions and how they propagate across the Solar System.

The primary energy source for space weather is the rotation of the Sun, though the natural fusion reactor in the core of the Sun also plays an essential role. What brings these together is their roles in creating the Sun's magnetic field. Heat from the core is transported to the surface by convection; this drives dynamo processes in the electrically conducting plasma that fills the outer layers of the Sun, resulting in the generation of magnetic fields. These magnetic fields gradually rise to the surface of the Sun and emerge into the Sun's atmosphere, known as the corona. Sunspots are then seen where strong magnetic fields thread the solar surface. This simple picture is then made more complex by the differential rotation of the Sun. Different latitudes of the Sun's surface rotate at different speeds, being faster at the equator and slower at high latitudes. Thus, as a loop of magnetic field rises to and through the solar surface, it will be stretched out as those parts of the loop at lower latitudes move ahead of those at higher latitudes. The overall effect is to distort the magnetic fields produced by convection, so that they develop a complex set of magnetic topologies. The creation of these topologies converts solar rotational energy into magnetic energy stored in complex magnetic structures in the solar corona. This sets the scene for the Sun to produce the phenomena that bring space weather to Earth.

3.6.1 Coronal Mass Ejections

The most significant of these phenomena are ejections of coronal plasma into interplanetary space, what we call coronal mass ejections or CMEs. The Sun's corona is very hot and would naturally expand into interplanetary space (Parker 1958) unless constrained by the magnetic fields which permeate the corona. However, these fields are not stable and can break down due to magnetic reconnection. This plasma process can reduce the complexity of the magnetic structures in the corona and may lead to some structures becoming magnetically detached from the Sun. Thus these structures, and the plasma embedded in them, may escape into interplanetary space accelerated in part by conversion of magnetic energy into kinetic energy.

As a result the Sun frequently (several times a day) ejects clouds of plasma into interplanetary space where we observe them as CMEs. CMEs vary enormously—in their origins on the Sun and in their size and speed. For example, some CMEs have their origin in the magnetic structures associated with active regions on the solar surface, that is, regions of intense magnetic fields and sunspots, but others may arise from the large magnetically confined loops of plasma which rise high above the solar surface (known as prominences when seen against space above the limb of the Sun, and filaments when seen in front of the solar disc). This variety of sources suggests a variety of generation mechanisms; however, that remains a topic for research.

The size, and in particular the speed, of CMEs are key factors in assessing what impacts will arise when they arrive on Earth, as they prescribe the amount of energy being carried by the CME. But how much of that energy will enter the Earth system? This is determined by two factors. Most obvious is whether the CME will actually hit the Earth—CMEs have a limited extent and can easily miss the Earth. We will see a significant impact only if the main part of a CME hits the Earth. It's perhaps useful to draw an analogy where the ejection of CMEs by the Sun is compared to a drunk throwing punches randomly—the punch is potentially dangerous, but we need to watch out and take precautions only if it comes directly toward us.

But, even if a CME hits the Earth, we may be shielded by the Earth's magnetic field. This deflects the hot plasma flowing out of the Sun, both the continuous background flow, which we call the solar wind, and any enhancements such as CMEs. The geomagnetic field blocks the flow creating a diamagnetic cavity around the Earth. This is the magnetosphere—the region of space dominated by the Earth's magnetic field. However, this magnetic barrier can be broken down by magnetic reconnection. Reconnection will occur when adjacent magnetic fields have opposite directions. Thus, energy from a CME will enter the magnetosphere when the CME magnetic field is southward, opposite to the northward field in the Earth's magnetosphere. This inflow drives what we call the Dungey or substorm cycle. Magnetic flux is transported from the dayside to the nightside of the Earth, increasing the amount of magnetic flux in the tail of the magnetosphere and driving the cross-polar flow in the polar ionosphere as discussed above. At some point, typically after 1 to 3 hours, the build-up of magnetic flux in the tail will breakdown via magnetic reconnection, explosively converting magnetic energy into kinetic energy. Some of that energy release will be directed toward the Earth producing aurora and ionospheric electric currents. If the energy inflow from the solar wind continues for many hours it will drive a repeated cycle of accumulation of magnetic energy in the tail followed by explosive release. This is the substorm cycle, the fundamental dynamical cycle of Earth's magnetosphere.

Thus, coronal mass ejections can drive substorms leading to a wide range of space weather impacts that arise from geomagnetic activity, as discussed above. These include, most notably, the geomagnetically induced currents which can disrupt power grids, rail systems, and pipeline operations. They also have a major influence on the ionosphere and thermosphere, impacting many radio technologies as well as changing satellite orbits. They can also enhance high-energy electron fluxes in the outer magnetosphere, leading to deep dielectric charging of satellites.

Thus, CMEs have a broad range of space weather impacts on infrastructures critical to energy distribution, transport, and communications. This, together with their ability to generate the most intense substorms, makes CMEs undoubtedly the most dangerous of space weather phenomena. As a result, they have been, and continue to be, a major focus for research. The key focus today is the development and refinement of techniques for observing CMEs, for modeling their propagation to Earth and for the generation of forecasts of their arrival and likely impact on Earth. Several space weather services, such as the U.S. Space Weather Prediction Center and the United Kingdom's Met Office Space Weather Operations Centre, are already providing forecasts of CME arrival based on state-of-the-art observations and models. But there is much interest in expanding these forecasts through better data acquisition and modeling. This is driving work to develop new measurement programs in the United

States (Biesecker and Socker 2017) and Europe (Luntama et al. 2017a), as well as research to deliver better science and models (Schrijver et al. 2015).

3.6.2 Stream Interaction Regions

The Sun also emits more steady plasma outflows, which complement the bursty outflows of CMEs. They may roughly be characterized as low-speed streams from the equatorial regions of the Sun and high-speed streams from its polar regions. Low and high must, of course, be understood in context: low speed is 300 to 400 km s^{-1}, while high is 600 to 700 km s^{-1}. These are very high speeds in human terms, but quite natural in terms of the plasma physics at work in the corona.

These streams combine to form the background solar wind which fills all interplanetary space. This will be compressed where the two streams meet such that the high-speed wind catches up on the low-speed wind (as shown in Figure 3.5). This is a fairly common occurrence because the boundary

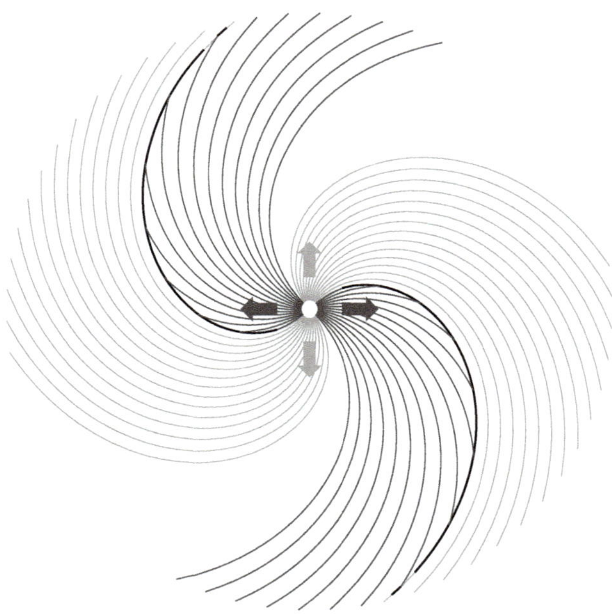

FIGURE 3.5 Schematic showing four solar wind streams leaving the Sun at low latitudes, each separated by 90° in longitude: two high-speed streams marked by the dark arrows, and two low-speed streams marked by lighter arrows. The streams flow radially away from the Sun, so the magnetic fields in the streams exhibit spiral patterns as shown here. An SIR (bold curve) forms where each high-speed stream catches up with the preceding slow-speed stream. (Figure by the author.)

between the high-speed and low-speed streams varies in latitude. Thus, as the Sun rotates, a low-speed stream emitted in one direction will be followed by a high-speed stream emitted in the same direction. This will inevitably lead to a region of compressed solar wind known as a stream interaction region (SIR). The Sun sometimes exhibits a very stable pattern of low- and high-speed streams lasting over two or more solar rotations. When this happens the SIRs will also exhibit a stable pattern which corotates with the Sun; hence, these SIRs are also referred to as corotating interaction regions (CIRs).

The compression in an SIR means that the plasma in the SIR has a higher energy density than the background solar wind. Thus, when an SIR passes over the Earth it can drive enhanced geomagnetic activity leading to substorms just as CMEs do, and it may be sustained by the high-speed solar wind that follows the SIR. In practical terms SIRs have similar impacts to those of CMEs. Thus, it is vital that space weather forecasting services include the detection and forecasting of SIRs as well as CMEs. However, this is generally more straightforward since SIRs are quasistable structures. Their likely existence can be inferred from observations of magnetic fields near the solar surface—looking for regions where those magnetic fields enable easy escape of plasma into interplanetary space. Additional information can come from solar wind monitors in the regions of interplanetary space to the east of the Sun, which will be swept by an SIR several days before it encounters the Earth. The Lagrange L5 point lies in this region and there is growing interest in making operational measurements at this point (Hapgood 2017b).

3.6.3 Solar Energetic Particles (Solar Radiation Storms)

Solar energetic particle events (SEPs), or more colloquially solar radiation storms, are another important solar emission. They typically last from many hours to a few days, though in severe conditions a series of events can merge to produce longer periods of elevated radiation levels—as happened in October 1989 (see Figure 3.1). Multiday events are thought to arise from the shock waves that form ahead of fast CMEs as they travel away from the Sun (Reames 1999). CMEs with speeds above about 800 km s^{-1} will generate such shocks as they plough through the undisturbed plasma of the solar wind—analogous to the way that supersonic aircraft generate shock waves in Earth's atmosphere. The shock ahead of a CME will persist throughout its journey to and beyond Earth, so it naturally produces energetic particles during the several days of this journey.

Short (<24 hours) SEP events are thought to have a different source—the reconnection processes in the solar corona (Drake et al. 2009), which also produce solar flares (see the next section). In this case, the source persists for a much shorter duration than a CME shock, and the energetic particles will escape into interplanetary space only if they can reach magnetic field lines open to interplanetary space. This combination of short source duration and variable access to interplanetary space provides a good explanation for short ("abrupt") SEP events, while explaining why we observe many reconnection events (as evidenced by solar flares) not associated with SEP events seen at Earth.

The journey of SEPs from the Sun to the Earth is guided by the interplanetary magnetic field, which usually takes a spiral pattern (Parker 1958) as shown in Figure 3.5. As a result, the near-Earth signature of an SEP event is dependent of the location of the source region, specifically its solar longitude relative to the central meridian as seen from Earth (Reames 1999). A source on the west side of the Sun will have good magnetic connectivity to the Earth and hence a significant signature will be seen near Earth a few tens of minutes after the event starts (the particles travel at a substantial fraction of the speed of light). Particles from CME sources near the central meridian or even on the east side of the Sun are likely to reach the Earth—with some delay. The CME may distort the interplanetary magnetic field sufficiently to direct particles close to Earth, especially when combined with cross-field scattering of the SEPs.

SEPs can greatly increase (several orders of magnitude) the radiation environment in near Earth space as shown in Figure 3.1. This leads to a wide range of radiation effects as discussed previously—most obviously radiation damage and single event effects (SEEs) on satellites. If the energy spectrum of the SEP event includes significant fluxes above 400 MeV, it can enhance the atmospheric radiation environment producing a greater risk of SEEs in aircraft and ground systems, and causing increases in radiation doses for aircrew and aircraft passengers. These events can also force a closure of the Arctic airspace, thereby disrupting air traffic between North America and Asia.

The main focus for mitigation of SEP events is better engineering to make systems more resilient to SEP impacts, and nowcasting to confirm that an event is in progress, so that operators can recognize and address problems arising from SEP impacts. There is also significant demand for all-clear forecasts, that is, forecasts that an event will not occur in the next hour or next 6 hours. These are valuable to organizations planning critical operations, such as a satellite launch, to make sure they will not be disrupted by an SEP event as a failure could have catastrophic results (e.g., loss of a satellite launch).

3.6.4 Solar Flares

Reconnection processes in the solar corona are also thought to produce substantial fluxes of energetic electrons at keV to MeV energies. Many of these electrons will travel toward the solar surface where they collide with denser matter, thus producing an intense burst of electromagnetic radiation particularly in the form of X-rays and EUV. This is a solar flare—a bright flash on the surface of the Sun, sometimes followed by less intense emissions as the flare causes hot material to rise a little way above the solar surface. The flare may also produce emissions at optical and gamma-ray wavelengths.

The electromagnetic radiation from a flare will travel out into interplanetary space, filling all 2π of solid angle above the flare. This broad spray of radiation from a flare is more likely to reach Earth compared to the more focused punch provided by a CME, but conversely the flare radiation is much more dispersed, leading to a lesser impact on Earth. The flare will increase ionization rates in the F-region ionosphere (due to flare EUV) and the D-regions (due to X-rays). The consequence for the F-region is an increase in plasma density which smoothly varies across the dayside of the Earth (following the sine of the Sun's elevation in the sky) so that density changes measured by a network of sensors can easily be extrapolated to other locations. Thus, systems such as GPS can fully mitigate flare effects using near-real-time corrections from augmentation systems such as EGNOS. The consequence for the D-region is blackout of MHz communications on the dayside of the Earth. However, this will usually last only for a few tens of minutes and may be mitigated by use of alternate communications systems such as L-band satcom.

3.7 MANAGING SPACE WEATHER RISKS—THE IMPORTANCE OF POLICY

Space weather is now a topic of considerable interest for professional risk managers around the world, both in governments and industry. This interest is driven in major part by a growing realization that societies around the world need to manage risks better. There are many factors (natural hazards, accidents, and malicious human actions) that can severely disrupt our societies, and we should build up societal resilience to these factors. Space weather fits into this picture as a natural hazard.

Societies tend to be well prepared for major hazards which occur on annual or decadal timescales, and therefore are well embedded in people's personal experience. Major hazards that repeat on centennial timescales are more challenging to prepare for. These are frequent enough to be a serious concern, but generally lie beyond personal or organizational experience. Scientific knowledge of the hazard is vital to provide robust evidence to underpin good policy decisions. A good example of this is the 2010 eruption of the Eyjafjallajökull volcano in Iceland. The ash released by this eruption caused major disruption of air transport in northern Europe. It was a scenario that had been anticipated by experts in the scientific community (Schmidt 2015) but which had not then been appreciated by risk managers in governments and the aviation industry. The subsequent catch-up of understanding means that the next event (and we can be sure that Icelandic volcanoes will erupt huge ash clouds in the future) will be much better managed. One of the broader lessons learned from the Eyjafjallajökull eruption was the need to use scientific knowledge to identify and assess major hazards.

Space weather is one of the hazards highlighted as a result of this lesson. The scientific community already had a substantial understanding that the great geomagnetic storm of September 2–3, 1859, the Carrington Event, could provide an example of a 1-in-100-year event (Clauer and Siscoe 2006; Space Studies Board 2009). That space weather event was initiated by an intense event on the Sun on September 1, 1859 during which a huge solar flare was observed on the Sun (Carrington 1859; Hodgson 1859). This occurred simultaneously with a large magnetic perturbation observed in London (Stewart 1861) arising because extra ionization from flare X-rays increased ionospheric conductivity. We can also infer that a fast CME from the Sun reached Earth in only 17 hours (a mean speed of 2400 km s^{-1}) and caused a huge magnetic storm seen around the world (Stewart 1861; Ptitsyna et al. 2012; Nevanlinna 2008; Tsurutani et al. 2003).

The Carrington Event is well documented in the scientific literature of the time and in modern reviews of that literature. This literature, together with surviving measurements and other records, has been widely exploited to assess how a repeat of the Carrington Event would severely disrupt modern societies (Cannon et al. 2013; Hapgood et al. 2016; National Science and Technology Council 2017). They show that a repeat could cause serious societal disruption via impacts on many technologies including power grids, transport systems, satellite navigation, and communications systems. That message has been reinforced by studies of more recent space weather events, notably the great geomagnetic storm of May 1921 (Karsberg et al. 1959; Kappenman 2006), and the high-speed coronal mass ejection that erupted on the solar farside on July 23, 2012. Simulations of the 2012 event (Baker et al. 2013) suggest it would have caused severe disruptions had it hit Earth. Fortunately, that one missed us.

But this luck will not last forever; one day there will be a space weather event which has the potential to severely disrupt everyday life. This needs to be considered by policy makers alongside other major natural hazards. For example, in the United Kingdom, severe space weather is now included in the national risk register (Cabinet Office 2015). This is a high-level assessment of various risks facing the United Kingdom and is shown in Table 3.1. This matrix looks at both the potential impact on everyday life—scored on a logarithmic scale from limited to catastrophic—and the occurrence probability as derived from scientific studies. The potential impact is assessed in terms of several factors including: (1) disruption to daily life, (2) disruption of the economy, and (3) psychological problems.

TABLE 3.1 Risk Matrix for the 2015 UK National Risk Assessment

Impact					
Catastrophic (5)				Pandemic influenza	
Severe (4)			Coastal flooding Widespread electricity failure		
Moderate (3)		Major transport accidents Major industrial accidents	Effusive volcanic eruption Emerging infectious diseases Inland flooding	Severe space weather Low temperatures and heavy snow Heat waves Poor air quality events	
Minor (2)		Public disorder Severe wildfires	Animal diseases Drought	Explosive volcanic eruption Storms and gales	
Limited (1)			Disruptive industrial action		
Probability in next 5 yrs	1-in-20,000 to 1-in-2000	1-in-2000 to 1-in-200	1-in 200 to 1-in-20	1-in-20 to 1-in-2	> 1-in-2

Source: Cabinet Office, 2015. *National Risk Register of Civil Emergencies*. https://www.gov.uk/government/collections/national-risk-register-of-civil-emergencies (accessed March 8, 2017).

The position of severe space weather in the matrix reflects the UK Government's assessment of the potential impact and the scientific community's assessment that a Carrington-class severe space weather event is likely to occur at least once in a 100 years. The position of risks in the matrix is a guide to their importance, and hence on how to prioritize resources toward their mitigation, with risks located toward the upper right clearly being the most significant. But it is important to recognize that positions in the matrix must not be taken as an absolute ranking—statements giving a numerical ranking of risks are unhelpful. The matrix is a guide, not a competition.

Policy makers will use this guide to explore how to mitigate each risk. In particular, they will assess: (1) how much the various risks can be managed using generic capabilities, for example, outages of power grids can be addressed through well-established "blackstart" procedures; and (2) where does each risk require specific capabilities, for example, access to forecasts and nowcasts in the case of space weather. Given these assessments policy makers can then consider how best to invest time and money in risk mitigation, for example, will the investment significantly improve the security of our societies? Will the investment deliver overall cost savings that are significantly larger than the cost of investment? As a rule of thumb public investment is justified when the savings are at least three times the cost of mitigation. Thus, there is an emerging field of socioeconomic studies on the impact of space weather, enabling space weather experts to work with economists.

The space weather impact on power grids has received by far the most attention in economic studies. A number of reports (Space Studies Board 2009; Lloyd's 2013) suggested that transformer damage during a severe space weather event could lead to months or years of power outages in the United States and an economic impact amounting to trillions of dollars. A recent study by Oughton et al. (2017) looked at a range of scenarios including this very high impact scenario as the upper limit, but also showed that the impact was reduced as grid and transformer resilience improves. The very high impact scenario has been widely challenged in other studies (NERC 2012; Cannon et al. 2013), which suggested that the likely impact of a severe event in several countries was regional voltage collapse with limited transformer damage, leading to power outages of a few days for most of those affected. Some areas would experience longer outages, but such outages could be handled by generic procedures already in place to deal with damage from terrestrial weather events. This lesser scenario is still a severe impact on all concerned, but with an economic impact of billions of dollars rather than trillions. The difference between these scenarios is still unresolved and policy development needs to be underpinned by further research on a number of key issues including: (1) the size of the area impacted by severe space weather (Pulkkinen et al. 2015); and (2) the need to consider space weather impacts on transformers as an ageing process (Gaunt 2014).

Most studies of the economic impact of space weather have so far focused on the total costs arising from that impact. Such studies are of great value for the insurance industry since it is concerned with the total value at risk from space weather, and wishes to determine a price at which it can reasonably offer insurance against this risk. However, it is not what is required for most policy work by governments and industry. What they require is an assessment of how much the costs of the impact will be reduced by implementing specific measures to mitigate the risk (e.g., improved forecasts), and conversely how much the costs will increase if existing capabilities are not maintained and renewed (e.g., replacement of critical space weather monitoring instruments). A UK study (Biffis and Burnett 2017) suggested that current forecasts based on data from instruments operating at the L1 point substantially reduce costs (4 billion pounds for the United Kingdom alone) arising from impact of a severe event on power grids and aviation. The study also showed that enhanced forecasts, adding data from instruments operating at the Lagrange L5 point, would bring further benefit (2 billion pounds for the United Kingdom alone) for both sectors. A parallel studied funded by ESA (Luntama et al. 2017b) broadly supported this conclusion, and also showed further significant benefits could accrue by mitigating impacts on resource exploitation (e.g., oil and gas) and on road logistics.

In summary, scientific understanding is enabling policy makers to develop plans to handle adverse space weather. In particular, the science of space weather shows that substantial risks arise from occasional severe events at the 1-in-100-year level appropriate to national risk management plans. As a result space weather is being included in the risk plans of many countries around the world. This is stimulating many actions to raise awareness of the challenges that space weather poses for industry and all levels of government, and thus improves society's resilience to space weather.

3.8 LOOKING TO THE FUTURE

The risks posed by space weather are a consequence of our societal dependence on advanced technologies. The phenomena that cause space weather have existed ever since the Sun and the Earth formed 5 billion years ago, but had no adverse impacts on human activities until the middle of the nineteenth century, when the electric telegraph became the first technology to be affected by what we now call space weather (Barlow 1849). This was followed by a slow growth of impacts on other technologies including telephones (Preece 1894), transoceanic cables (New York Times 1921), long distance radio communications (Anderson 1928), and electric power transmission (McNish 1940). Until the 1960s, these impacts had only minor societal consequences, since everyday life did not then rely heavily on electric power and electronic communications. But electric power is now absolutely fundamental to everyday life, and we now take for granted, both at home and at work, that we can communicate almost instantly to the four corners of the Earth. In addition, a raft of new technologies have revolutionized our lives: airliners allow us to travel quickly and safely to distant places at reasonable cost; electronic banking services such as debit and credit cards have revolutionized how money circulates in the economy; and satellites provide a range of critical services including accurate timing, precise location and navigation, and communications. These technological advances have greatly improved the standard of living for billions of people around the planet, but they have also created a vulnerability to space weather. Thus, as new technologies emerge, we need to assess how they change the risks space weather poses for modern societies.

It is likely that some of those changes will be benign. For example, the growing use of frequencies above 4 GHz for satellite communications with aircraft (e.g., to support good quality internet links) will, as a side effect, gradually reduce the risk that space weather disrupts communications between pilots and aircraft traffic control centers. However, given that existing aircraft may have two decades of useful life, it maybe several decades before this space weather risk to aircraft communications is phased out. Even then there may be some desire to retain MHz radio communications as a backup not dependent on satellites and therefore not at risk from space weather impacts on satellites. Another example of change is the evolution of electric power grids as low carbon sources begin to dominate the generation of electricity. This is changing the topologies, and modes of operation, of power grids as many low carbon sources are located far from consumers and some (e.g., solar, wind, and tidal) have outputs that are naturally variable in time. Thus, grid operators have to understand how GIC will interact with these complex power flows.

Perhaps the most important example of change is the embedding of GNSS technologies in an ever-growing range of applications, which exploit its ability to provide precise location and time. These applications are now a major focus for commercial development and are producing some valuable services. But this needs to be balanced by assessments of how these applications will function in adverse space weather conditions, especially when those conditions deny access to GNSS signals for periods of hours or even a few days (Cannon et al. 2013). A simple example is the use of GNSS to direct drivers of road vehicles to unfamiliar locations, a capability needed by many logistical services. A recent study funded by ESA noted the growing risk of impact on logistical services as they become dependent on GNSS and estimated the value at risk, over the period 2017–2032, to be in excess of 3 billion Euros (Luntama et al. 2017b). GNSS is now being promoted widely (and rightly) as a technology that can drive innovation in many commercial services, but this drive needs to be balanced by an awareness of the vulnerability of GNSS to disruption by a number of phenomena, not least space weather (Curry 2014; Hapgood 2017a).

Finally, an important issue for the future is the impact of space weather on the emerging technology of driverless cars. This is a technology with the potential to improve both the safety and the experience of road travel, a potential widely recognized by industries and governments around the world. But it is also a technology where failure can have lethal consequences—as sadly demonstrated by recent accidents (Stilgoe 2017), while remaining a technology with several components vulnerable to space weather (e.g., digital devices [Wood and Caustin 2006] and GNSS [Hapgood 2017a]). Space weather should be incorporated in the risk scenarios to be addressed as part of the development of driverless cars. It is not clear that this is yet the case, so the space weather expert community needs to raise awareness of the issues.

The advent of driverless cars provides a valuable model for future work in space weather. As new technological systems emerge, we need to look at what technologies are embedded inside them and assess whether they create a system vulnerability to space weather. Such vigilance is an essential complement to technological innovation, which we now realize is a key driver of economic development.

REFERENCES

Air University, 2003. *Space Primer: Space Environment*, Chapter 6, pp. 6–13. http://space.au.af.mil/primer/space_environment.pdf (accessed June 27, 2017).

Anderson C.N., 1928. Correlation of long wave transatlantic radio transmission with other factors affected by solar activity. *Proc Inst Radio Eng*, 16, 297–347.

Aplin K.L. and R.G. Harrison, 2014. Atmospheric electric fields during the Carrington flare. *Astron Geophys*, 55(5), 32–37.

ATSB (Australian Transport Safety Bureau), 2011. In-flight upset, 154 km west of Learmonth, Western Australia, October 7, 2008, VH-QPA, Airbus A330-303. Report AO-2008-070. ISBN 978-1-74251-231-0.

Avanti, 2017. SBAS-AFRICA. http://sbas-africa.avantiplc.com/ (accessed April 1, 2017).

Axford W.I. and C.O. Hines, 1961. A unifying theory of high-latitude geophysical phenomena and geomagnetic storms. *Can J Phys*, 39, 1433–1464.

Baker D.N. et al., 1994. Relativistic electron acceleration and decay timescales in the inner and outer radiation belts: SAMPEX. *Geophys Res Lett*, 2, 409–412.

Baker D.N. et al., 2013. A major solar eruptive event in July 2012: Defining extreme space weather scenarios. *Space Weather*, 11, 585–591.

Barlow W.H., 1849. On the spontaneous electrical currents observed in wires of the electric telegraph. *Phil Trans R Soc Lond*, 139, 61–72.

Biesecker D.A. and D. Socker, 2017. *NOAA L1 Status and Studies*. https://www.ukssdc.ac.uk/meetings/L5InTandemWithL1/talks/session02/08_Biesecker%20-%20NOAA%20L1%20Status%20and%20Studies.pptx (accessed June 24, 2017).

Biffis E. and C. Burnett, 2017. *IPSP Space Weather Socio-Economic Study*. https://www.ukssdc.ac.uk/meetings/L5InTandemWithL1/talks/session01/02_L1-5%20-%20IPSP%20PresentationEBv2.pptx (accessed June, 24 2017).

Board, Space Studies, and National Research Council, 2009. *Severe Space Weather Events—Understanding Societal and Economic Impacts: A Workshop Report*. National Academies Press.

Bolduc L., 2002. GIC observations and studies in the Hydro-Québec power system. *J Atmos Sol Terr Phys*, 64, 1793–1802.

Cabinet Office, 2015. *National Risk Register of Civil Emergencies*. https://www.gov.uk/government/collections/national-risk-register-of-civil-emergencies (accessed March 8, 2017).

Cannon P. et al., 2013. *Extreme Space Weather: Impacts on Engineered Systems and Infrastructure*. UK Royal Academy of Engineering, London. ISBN 1-903496-95-0.

Carrington R.C., 1859. Description of a Singular Appearance seen in the Sun on September 1. *Mon Not R Astron Soc*, 20, 13–15.

CDC (Centers for Disease Control and Prevention), 2015. *ALARA—As Low As Reasonably Achievable*. https://www.cdc.gov/nceh/radiation/alara.html (accessed April 24, 2017).

Clauer C.R. and G. Siscoe, 2006. The Great Historical Geomagnetic Storm of 1859: A modern look. *Adv Space Res*, 38, 115–388.

Curry C., 2014. *SENTINEL Project: Report on GNSS Vulnerabilities*. Chronos Technology Ltd. http://www.chronos.co.uk/files/pdfs/gps/SENTINEL_Project_Report.pdf (accessed April 28, 2017).

Drake J.F. et al., 2009. A magnetic reconnection mechanism for ion acceleration and abundance enhancements in impulsive flares. *Astrophys J Lett*, 700, L16.

Dyer C.S. et al., 2007. Solar particle events in the QinetiQ atmospheric radiation model. *IEEE Trans Nucl Sci*, 54, 1071–1075.

Erinmez I.A. et al., 2002. Management of the geomagnetically induced current risks on the national grid company's electric power transmission system. *J Atmos Sol Terr Phys*, 64, 743–756.

Eroshenko E.A. et al., 2010. Effects of strong geomagnetic storms on northern railways in Russia. *Adv Space Res*, 46, 1102–1110.

Gaunt C.T., 2014. Reducing uncertainty—Responses for electricity utilities to severe solar storms. *J Space Weather Space Clim*, 4, A01.

Gaunt C.T. and G. Coetzee, 2007. Transformer failures in regions incorrectly considered to have low GIC-risk. *Proceedings of the IEEE Powertech Conference*, July 2007, Lausanne, Switzerland.

Gauthier L. et al., 2001. EGNOS: The first step in Europe's contribution to the global navigation satellite system. *ESA Bulletin*, 105, 35–42. http://esamultimedia.esa.int/multimedia/publications/ESA-Bulletin-105/.

Gold T. and D.R. Palmer, 1956. The solar outburst, February 23, 1956—Observations by the Royal Greenwich Observatory. *J Atmos Terr Phys*, 8, 287–290.

Gurnett D.A., 1974. The Earth as a radio source: Terrestrial kilometric radiation. *J Geophys Res*, 79, 4227–4238.

Hambling D., 2014. Burnout. *New Scientist*, 223, 42–45.

Hapgood M., 2017a. Satellite navigation—Amazing technology but insidious risk: Why everyone needs to understand space weather. *Space Weather*, 15, 545–548.

Hapgood M., 2017b. L1L5Together: Report of workshop on future missions to monitor space weather on the sun and in the solar wind using both the L1 and L5 Lagrange Points as valuable viewpoints. *Space Weather*, 15, 654–657.

Hapgood M. and A. Thomson, 2010. *Space Weather: Its Impact on Earth and Implications for Business*. Lloyd's 360 Risk Insight. London, UK.

Hapgood M. et al., 2016. *Summary of Space Weather Worst-Case Environments*. Revised edition. RAL Technical Report RAL-TR-2016-06. http://purl.org/net/epubs/work/25015281 (accessed March 6, 2017).

Hey J.S., 1946. Solar Radiations in the 4–6 metre radio wave-length band. *Nature*, 157, 47–48.

Hodgson R., 1859. On a curious Appearance seen in the Sun. *Mon Not R Astron Soc*, 20, 15–16.

Horne R. et al., 2005. Wave acceleration of electrons in the Van Allen radiation belts. *Nature*, 437, 227–230.

HSE (Health and Safety Executive), 1992. *The Tolerability of Risk from Nuclear Power Stations*. http://www.onr.org.uk/documents/tolerability.pdf (accessed June 23, 2017).

ICRP, 1991. 1990 Recommendations of the International Commission on Radiological Protection. ICRP Publication 60. Ann. ICRP 21.

Kappenman J.G., 2006. Great geomagnetic storms and extreme impulsive geomagnetic field disturbance events—An analysis of observational evidence including the great storm of May 1921. *Adv Space Res*, 38, 188–199.

Karsberg A. et al., 1959. *The Influences of Earth Magnetic Currents on Telecommunication Lines*. Tele (English ed.), Televerket, Stockholm, 1–21.

Krausmann E. et al., 2015. *Space Weather and Rail: Findings and Outlook*. EU Joint Research Centre Report 98155.

Liu L. et al., 2016. Analysis of the monitoring data of geomagnetic storm interference in the electrification system of a high-speed railway. *Space Weather*, 14, 754–763.

Lloyd's, 2013. *Solar Storm Risk to the North American Electric Grid*. http://www.aer.com//sites/default/files/Solar_Storm_Risk_to_the_North_American_Electric_Grid_0.pdf (accessed June 26, 2017).

Lockwood M., 2007. Fly me to the Moon? *Nat Phys*, 3, 669–671.

Loh R. et al., 1995. The U.S. Wide-Area Augmentation System (WAAS). *Navigation*, 42, 435–465.

Luntama J. et al., 2017a. *ESA SSA Roadmap for the L5 Mission*. https://www.ukssdc.ac.uk/meetings/L5InTandemWithL1/talks/session02/05_SSA%20L5%20Roadmap.pptx (accessed June 24, 2017).

Luntama J. et al., 2017b. *Report on the ESA Space-Weather Socio-Economic Study*. https://www.ukssdc.ac.uk/meetings/L5InTandemWithL1/talks/session01/04_SSA%20SWE%20CBA%20Study.pptx (accessed June 24, 2017).

Marsden P.L. et al., 1956. Variation of cosmic-ray nucleon intensity during the disturbance of February 23, 1956. *J Atmos Terr Phys*, 8, 278–281.

McIlwain C.E., 1961. Coordinates for mapping the distribution of magnetically trapped particles. *J Geophys Res*, 66, 3681–3691.

McNish A.G.N., 1940. The magnetic storm of March 24, 1940. *Terr Magn Atmos Electr*, 45, 359–364.

Meredith N.P. et al., 2015. Extreme relativistic electron fluxes at geosynchronous orbit: Analysis of GOES E > 2 MeV electrons. *Space Weather*, 13, 170–184.

Miyake F. et al., 2012. A signature of cosmic-ray increase in AD 774–775 from tree rings in Japan. *Nature*, 486, 240–242.

Miyake, S. et al., 2017. Cosmic ray modulation and radiation dose of aircrews during the solar cycle 24/25. *Space Weather*, 15, 589–605.

National Science and Technology Council, 2017. *Space Weather Phase 1 Benchmarks*. http://www.ofcm.gov/publications/spacewx/DRAFT_SWx_Phase_1_Benchmarks.pdf (accessed March 8, 2017).

NERC (North American Electrical Reliability Corporation), 2012. *GMD Task Force Report: Effects of GMD on the Bulk Power System*. http://www.nerc.com/pa/Stand/Geomagnetic%20Disturbance%20Resources%20DL/2012_GMD_Report_112012.pdf (accessed June 24, 2017).

Nevanlinna, H., 2008. On geomagnetic variations during the August–September storms of 1859. *Adv Space Res*, 42, 171–180.

New York Times, 1921. *Cables Damaged by Sunspot Aurora*. May 17, 1921. http://query.nytimes.com/mem/archive-free/pdf?res=9407E2D61E3FEE3ABC4F52DFB366838A639EDE (accessed May 12, 2017).

NOAA, 2011. *NOAA Space Weather Scales*. http://www.swpc.noaa.gov/noaa-scales-explanation (accessed June 24, 2017).

Normand E., 1996. Single event upset at ground level. *IEEE Trans Nucl Sci*, 43, 2742–2750.

Normand E., 1997. Neutron-induced single event burnout in high voltage electronics. *IEEE Trans Nucl Sci*, 44, 2358–2366.

Oughton E.J. et al., 2017. Quantifying the daily economic impact of extreme space weather due to failure in electricity transmission infrastructure. *Space Weather*, 15, 65–83.

Parker E.N., 1958. Dynamics of the interplanetary gas and magnetic fields. *Astrophys J*, 128, 664.

Paulikas G.A. and J.B. Blake, 1979. Effects of the solar wind on magnetospheric dynamics: Energetic electrons at the synchronous orbit, in *Quantitative Modeling of Magnetospheric Processes* (ed. W.P. Olson), American Geophysical Union, Washington, DC, pp. 180–202.

Pisacane V.L., 2016. *The Space Environment and Its Effects on Space Systems*. Second Edition. AIAA, Reston, VA.

PourEVA, 2003. *Le Ministre DEWAEL reconnait la faillibilité du vote électronique grâce à un rayon cosmique complice!* http://www.poureva.be/article.php3?id_article=36 (accessed May 8, 2017).

Preece W.H., 1894. Earth Currents. *Nature*, 49, 554.

Ptitsyna N.G. et al., 2012. New data on the giant September 1859 magnetic storm: An analysis of Italian and Russian historic observations. *Proc. 9th Intl. Conf. "Problems of Geocosmos"*, October 8–12, 2012, St. Petersburg, Russia.

Public Health England, 2011. *Ionising Radiation: Dose Comparisons*. https://www.gov.uk/government/publications/ionising-radiation-dose-comparisons (accessed June 23, 2017).

Public Health England, 2014. *What Is Radon?* http://www.ukradon.org/information/whatisradon (accessed June 22, 2017).

Pulkkinen A. et al., 2015. Regional-scale high-latitude extreme geoelectric fields pertaining to geomagnetically induced currents. *Earth, Planets and Space*, 67, 93.

Pulkkinen A. et al., 2017. Geomagnetically induced currents: Science, engineering, and applications readiness. *Space Weather*, 15, 828–856.

Rao K.N., 2007. GAGAN—The Indian satellite based augmentation system. *Indian J Radio Space Phys*, 36, 293–302. http://nopr.niscair.res.in/handle/123456789/4707.

Reames D.V., 1999. Particle acceleration at the Sun and in the heliosphere. *Space Sci Rev*, 90, 413.

Schmidt A., 2015. Volcanic gas and aerosol hazards from a future Laki-type eruption in Iceland, in *Volcanic Hazards, Risks and Disasters* (eds. J.F. Shroder and P. Papale), Elsevier, Boston, pp. 377–397. http://doi.org/10.1016/B978-0-12-396453-3.00015-0.

Schrijver C.J. et al., 2015. Understanding space weather to shield society: A global road map for 2015–2025 commissioned by COSPAR and ILWS. *Adv Space Res*, 55, 2745–2807.

Stewart B., 1861. On the great magnetic disturbance which extended from August 28 to September 7, 1859, as recorded by photography at the Kew Observatory. *Phil Trans R Soc Lond*, 151, 423–430.

STFC, 2017. *ChipIR: Instrument for Rapid Testing of Effects of High Energy Neutrons*. http://www.isis.stfc.ac.uk/instruments/chipir/chipir8471.html (accessed March 31, 2017).

Stilgoe J., 2017. Tesla crash report blames human error—This is a missed opportunity. *The Guardian*, January 21, 2017. https://www.theguardian.com/science/political-science/2017/jan/21/tesla-crash-report-blames-human-error-this-is-a-missed-opportunity (accessed April 28, 2017).
Thomas B.C. et al., 2013. Terrestrial effects of possible astrophysical sources of an AD 774-775 increase in 14C production. *Geophys Res Lett*, 40, 1237–1240.
Thomson, A. and J. Wild, 2010. When the lights go out…. *Astron Geophys*, 51(5), 23–24.
Trinity House, 2016. *Satellite Navigation Ground Based Augmentations*. https://www.trinityhouse.co.uk/dgps (accessed April 1, 2017).
Tsurutani B.T. et al., 2003. The extreme magnetic storm of September 1–2, 1859. *J Geophys Res*, 108(A7), 1268.
Usoskin I.G. et al., 2013. The AD775 cosmic event revisited: The Sun is to blame. *Astron Astrophys*, 552, L3.
Whipple E.C., 1981. Potentials of surfaces in space. *Rep Prog Phys*, 44, 1197.
Wik M. et al., 2009. Space weather events in July 1982 and October 2003 and the effects of geomagnetically induced currents on Swedish technical systems. *Ann Geophys*, 27, 1775–1787.
Wood J. and E. Caustin, 2006. *Timely Testing Avoids Cosmic Ray Damage to Critical Auto Electronics*. http://www.eetimes.com/document.asp?doc_id=1272752 (accessed April 28, 2017).

CHAPTER 4

Hazards and Habitability

Galactic Perspectives

Richard J. Wilman, Pratika Dayal,
and Martin J. Ward

CONTENTS

4.1 BACKGROUND AND MOTIVATION

Most discussions of space risks and their associated societal challenges, including those elsewhere in this volume, focus on the near-Earth environment or are more generally limited to the Solar System. Indeed, for most practical purposes it is often implicit that the Solar System is the only "space" with which we need be concerned. Discussions of natural cosmic hazards have accordingly focussed mainly on the threats intrinsic to this arena, either in the form of space weather originating from the Sun (see Chapter 3) or the impact hazard posed by the rocky debris left over from the formation of our planetary system (see Chapter 2). In both cases, the threats to Earth range from localized impacts and disturbances, up to globally catastrophic "doomsday" events. However, with political will and realistic improvements in surveillance and mitigation technology, most of their impacts can be substantially reduced or eliminated.

The aim of the present chapter is to cast the net over a much wider range of space and time to assess the threats arising from beyond the Solar System. These include the known astrophysical hazards in our own *Milky Way Galaxy,* and the somewhat more uncertain risks, on which we can only speculate, stemming from the possible existence of other forms of life and intelligence elsewhere in the Galaxy. At first sight, such a task may seem of little practical interest, of concern only to astronomers and science fiction writers, and to those interested in the external factors affecting the evolution of life on Earth. We shall see that significant astrophysical events originating in the wider Galaxy are expected to impact the Earth only on timescales of millions to billions of years; while they may have influenced terrestrial conditions in our geological past, looking forward in time the threat they pose is likely of no greater concern to humankind than the expected end of habitable conditions on Earth a few billion years from now, as the Sun exhausts its fuel supply and expands to become a red giant star.

We first review the risks and impacts from phenomena occurring in the wider Galaxy in so far as they affect the Earth and its environment. Potentially, at some point in the future our technology, and possibly humanity itself, will venture beyond the Solar System and will need to consider the in situ hazards of interstellar space. We will not explicitly concern ourselves with such far future challenges—only now, after four decades of interplanetary flight, are the Voyager probes reaching the edge of the Solar System. This situation may, however, conceivably change with ventures such as the Breakthrough Starshot initiative*—a mission to send a probe to the nearby star Alpha Centauri in under 20 years of travel time.

* https://www.breakthroughinitiatives.org.

Until very recently, our entire knowledge of our Galaxy and the Universe beyond relied on observations of electromagnetic radiation or energetic cosmic rays (the detection of the first gravitational waves by Abbott et al. (2016) has opened a new window on the cosmos, although this branch of astronomy is still in its infancy). Yet for all the wealth of information received by such means and the spectacular progress in astrophysics and cosmology thereby enabled over the last century, large telescopes, sophisticated detectors, and long exposure times are needed to detect the sources of interest or to study them in any detail. The possibility that radiation from such distant objects could pose a hazard to Earth may at first seem surprising, yet modern observational techniques—especially those operating in higher energy X-ray and gamma-ray wavebands—have brought to light a range of exotic astrophysical hazards which, under certain conditions, could inflict severe damage on the Earth and its biosphere. Most relate to the explosions of massive stars as supernovae or gamma-ray bursts. The absence of serious impacts of this nature in our recent recorded history is an observational selection effect, and cannot in and of itself be used to infer that the risk they pose is negligible. Although calculations indicate that such events are extremely rare—the probability is only significant on geological timescales—this does not justify neglecting them entirely. Indeed, if we consider risk as being the probability of occurrence of a particular event within a given timeframe (very small for timescales of interest to human civilization) multiplied by the consequent damage if the event occurs (enormous, for the complete destruction of the biosphere), the overall risk in this sense is not negligible.

Following on from stellar explosions, we then consider the risks posed by black holes as they accrete material, with an emphasis on the consequences of a flare-up of the currently dormant 4 million solar mass black hole, known as Sagitarrius A*, at the center of the Galaxy.

The hazards discussed so far are forms of *radiative transient*; other impacts arise from the dynamical effects of close encounters between the Solar System and passing stars and molecular clouds; such perturbations may disturb the cometary reservoir—the Oort Cloud—in the outer Solar System, sending swathes of potential planetary impactors into its inner regions (see Chapter 2), or in extremis, destabilize the orbital configuration of the planets themselves. Many of these dynamical risks are closely synchronized with the orbital path of the Sun on its 225 million journey around the Galaxy; on timescales of 10–100s of millions of years the Sun oscillates above and below the dense Galactic mid-plane, and in and out of its spiral arms. The concentration of star formation in the spiral arms may also induce secular variations in the dust and cosmic ray backgrounds in the Solar neighborhood, which in turn affect the terrestrial Solar flux and imprint signatures in the historical climate record.

We then abandon our geocentric perspective and consider how astrophysical hazards may affect the likelihood of life arising elsewhere in our Galaxy and others over the course of cosmic time. Interest in the issue of planetary *habitability*—the ability to support and even to give rise to life as we know it—has acquired new impetus with the discovery since the mid-1990s of several thousand planets around stars other than the Sun, a total which is rapidly increasing and that now includes planets comparable in mass to the Earth in orbit around Sun-like stars. Given a favorable combination of in situ planetary conditions and host star, external hazard considerations may dictate when and where in the Galaxy life can take hold and flourish.

For the Milky Way, we review the evidence for the existence of the so-called *Galactic Habitable Zone*, a zone of Galactic radius where both the necessary ingredients for life are in place and the hazard threat is tolerable. Although planets appear to be a common by-product of star formation, frustratingly little is known about them beyond their orbital parameters, masses, and/or radii and in a few cases, crude atmospheric properties; the rest is theory, conjecture, and speculation. However, the next generation of ground- and space-based telescope technology will begin to erode this ignorance by establishing the extent to which potentially *habitable* planets are in fact *inhabited*, in the sense of exhibiting detectable signs of ongoing biological activity in their atmospheres (Kaltenegger 2017). Until then at least, the possibility remains that such life is abundant in our Galactic neighborhood.

With the latter possibility in mind, in the final section of this chapter we look beyond purely astrophysical considerations to assess the risks and challenges surrounding our search for life beyond Earth, specifically of the intelligent variety. We review the possible channels through which the discoveries of extraterrestrial life may occur, including the search for extraterrestrial intelligence (SETI), and touch upon the challenges for our society, which may arise in the event of intelligent signals being discovered or contact established. Conversely, if continued and increasingly intensive searches fail to uncover any evidence for primitive or intelligent life elsewhere in the Galaxy, we consider some possible explanations for such null results.

As will already be apparent, this chapter is quite distinct from the main thrust of this volume. With our focus on natural hazards, which may eventuate only over very long timescales, and on the largely unknown risks arising from the presence (or indeed the absence) of intelligent life elsewhere in the Galaxy, we are far removed from most practical risk management concerns. Nevertheless, there may be factors beyond our current understanding ("unknown unknowns"), and as such any event of this nature would surely constitute the ultimate "black swan"—to use the metaphor popularized by Taleb (2007)—a high impact, existentially devastating event whose subjective probability was hitherto considered to be too low to be of concern.

4.2 REVIEW OF GALACTIC RISKS OF ASTROPHYSICAL ORIGIN

In this section we review the known Galactic astrophysical hazards to which the Earth is exposed. We begin by showing in Figure 4.1 the structure of the Milky Way Galaxy.*

FIGURE 4.1 Artist's impression of our Galaxy from plan and edge-on perspectives. The bulk of the stars reside in a disk about 30 kpc across and 1 kpc thick at its maximum extent, with most star formation occurring in the spiral arms. The Sun resides close to the mid-plane of the disk about halfway out from its nucleus, which hosts a dormant 4 million solar mass black hole, Sgr A*. (Copyright Left: NASA/JPL-Caltech; right: ESA; layout: ESA/ATG medialab.)

* In this chapter, we follow convention and measure distances in parsecs (pc), where 1 parsec = 3.26 light-years.

4.2.1 Stellar Death Throes: Supernovae

As already noted, the terrestrial impact posed by astronomical sources beyond the Solar System, let alone the hazard they pose, is usually entirely negligible. A recent rare exception to this occurred on December 27, 2004 when a Galactic soft gamma-ray repeater (SGR) known as SGR 1806-20 blasted the Earth with such an intense pulse of gamma-ray radiation (Hurley et al. 2005; Palmer et al. 2005) that it saturated most space-borne detectors and caused detectable space weather effects, with disturbances to the ionosphere (Inan et al. 2005) and magnetosphere (Mandea and Balasis 2005). The source of the burst was a star-quake on a highly magnetized neutron star (or *magnetar*) some 20 km in diameter. The burst was the most intense explosion witnessed by humans in our Galaxy since the supernova observed by Johannes Kepler in 1604, and in 0.2 seconds carried more energy than the Sun emits in 250,000 years (Hurley et al. 2005). Fortunately, the site of the burst was many kiloparsecs distant, but if it had been within a few light-years of the Sun the atmospheric impact would have been potentially serious or even catastrophic for life on Earth. There are no known magnetars currently within this danger zone (Kaspi and Beloborodov 2017), but other Galactic transients may pose similar or greater threats within somewhat larger spheres of influence, as we shall now see.

The Sun is currently about halfway through its 10 billion year lifespan as a "main sequence" star, powered by the nuclear fusion of hydrogen into helium nuclei at its core. As the supply of core hydrogen is gradually exhausted, the Sun's outer layers will slowly expand with an increase in its overall brightness and it will become a red giant star. This brightening alone will likely render the Earth uninhabitable within $\sim$1 Gyr, even if it is not ultimately swallowed by the Sun's expanding atmosphere. The Sun will eventually shed this atmosphere and the core will survive into cosmic posterity as a white dwarf star. While catastrophic for life on Earth, the death of the Sun is a long and drawn out affair which poses no risk to its Galactic neighborhood.

Essentially all stars that start out with more than about 8 times the mass of the Sun, and a minority of those with lower masses (mainly those in binary star systems), will end their lives in an altogether more dramatic manner as *supernovae*. Over a period of a few weeks, such cosmic explosions briefly outshine their entire host Galaxy, rendering them visible over significant cosmological distances. They come in two main variants: (1) *Type Ia*, in which the accretion of matter onto a white dwarf star in a binary star system pushes the white dwarf above the *Chandrasekhar Limit* (1.4 Solar masses) for support by the pressure of degenerate electrons (a form of quantum mechanical gas); and (2) *Type II*, or core-collapse supernovae, which occur when the chain of nuclear fusion reactions in the cores of stars of eight Solar masses and above reach the energetic limit, leading to rapid inward collapse of the outer layers of the star. Elementary particle reactions in the core ultimately halt the implosion, leading to a "bounce" and an outwardly propagating supernova shock wave.

The dangers from supernovae consist of high-energy electromagnetic radiation, cosmic rays, and radioactive nuclei. Comprehensive reviews of their impact have been provided by Dar (2008), Bonnet and Woltjer (2008), Korschinek (2017), and Beech (2011). Visually dramatic though they may be, the existential hazard posed by a supernova explosion is localized within the Galaxy and limited to a sphere of order 10 pc in radius (Gehrels et al. 2003). Within this canonical "kill zone," the flux of cosmic rays and high-energy radiation is capable of causing a mass extinction. Increased atmospheric ionization (or in extreme cases the removal of the entire atmosphere) leads to ozone destruction and a corresponding increase in the penetration of solar ultraviolet radiation to a level harmful to life (see e.g., Melott and Thomas 2011). Hydrodynamical simulations corroborate this by showing that a supernova at 10 pc would compress the heliopause—the conventional edge of the Solar System, where the solar wind holds off the flow of the interstellar medium—to a radius of 1 astronomical unit (AU),* down from $\sim$100 AU under current ambient conditions (Field et al. 2008).

The short pulse of electromagnetic radiation from the supernova is the first form of hazardous radiation to reach Earth and travels directly to us; the cosmic rays (mainly protons) are accelerated to near-light speed as the supernova remnant evolves (over timescales of $\sim$50,000 yr) and their transport to Earth is guided by the interstellar magnetic field, resulting in a delay with respect to the electromagnetic signal. Melott et al. (2017) have shown that with a favorable configuration of the interstellar magnetic field the lethal effects of the supernova can potentially extend to 50 pc or beyond. Radioactive isotopes from the supernova remnant can also contaminate the Earth from such distances. Indeed, thanks to pioneering interdisciplinary work in geoscience and astrophysics, much attention has focused on the detection of widespread radioactive ^{60}Fe in deep ocean

* 1 AU is the Earth–Sun distance, 1.5×10^{11} m.

crust material (Knie et al. 1999). Recent analysis shows that this signal is principally due to a pair of supernovae which occurred around ~100 pc from the Earth around 2 million years ago, with lesser contributions from multiple supernova events in this volume on a 10-million-year timescale (Breitschwerdt et al. 2016; Wallner et al. 2016). Such events may have created the so-called "Local Bubble" of hot gas within which the Solar System is immersed, and although located beyond the "kill zone," they may have caused moderate atmospheric damage, as evident in the fossil record around this time (Thomas et al. 2016; Melott et al. 2017). The radioactive material reached the Solar System by traveling on dust grains; while its distribution on the Earth is influenced by atmospheric and oceanic effects, the signal on the Moon would retain the directional information (Fry et al. 2016; Crawford 2017).

The supernova rate in the entire Milky Way Galaxy is a few per century, most of which are not observed from Earth due to Galactic dust (see Adams et al. 2013 for recent rate estimates). With a uniform rate of star formation per unit area within the Galactic disk out to 13 kpc (e.g., Kennicutt and Evans 2012), and ignoring disk thickness, one would expect a supernova within 100 pc once every ~ 50 yr*$(13{,}000/100)^2$ $\sim$ 1 million yr; and within 10 pc once every ~100 million years (as also estimated by Beech 2011). Taking into account the nonzero thickness of the Galactic plane of ~200 pc, estimates by Fry et al. (2015), scaled to a disk radius of 13 kpc, imply that supernovae within the 10 pc kill radius occur on average every ~700 million years.

These are global long-term average rates, assuming that supernovae occur randomly and uniformly throughout the Galactic disk. In reality, most stars form in groups or clusters containing hundreds to thousands of stars. Most disperse quickly into the Galactic field population but can still be compact (~pc-scale) when the high mass stars explode as Type II supernovae (Type Ia supernovae occur several Gyr after the formation of the stellar system). Hence, if the Earth passes within the 10 pc "kill zone" of one supernova it is likely to be in the firing line of multiple such events occurring in relatively quick succession. Taking this clustering into account, the time interval between supernova bombardment episodes (e.g., the multiple supernovae responsible for the ^{60}Fe detections events) will be longer than the mean time between individual events and roughly a once per Gyr occurrence. The star formation rate at the Sun's position in the Galaxy has been roughly constant for the last 4.6 Gyr, that is, since the birth of the Sun according to models by Naab and Ostriker (2006), implying that the local supernova rate has been approximately constant during this time (albeit modulated by the passage of the Sun through the Galaxy's spiral arms, where most star formation occurs).

So much for the long-term statistical risk. Of more interest is whether there are any known stars in the solar neighborhood likely to go supernova. As discussed by Korschinek (2017) following investigations by Beech (2011), the nearest Type Ia supernova candidate is the star IK Pegasi currently 46 pc from the Sun; it will reach its closest approach to the Sun at just under 39 pc in 1 million years from now and is expected to explode in 1.9 billion years time, by which time it will be too distant to pose any hazard to us. The nearest Type II supernova candidate is Betelgeuse in Orion, currently $\sim$200 pc away (and receding from us), and likely to explode sometime within the next 2 million years. Although much further away (2.3 kpc), eta Carinae, which underwent a dramatic long-term brightening in the mid-nineteenth century is another commonly cited supernova candidate (see Thomas et al. 2008).

4.2.2 Gamma-Ray Bursts

Some of the most massive stars, for reasons not yet fully understood, end their lives as *gamma-ray bursts* (GRBs) in which the explosive energy is channeled into highly directed, *relativistically beamed* jets, visible to about 1% of randomly located observers. Characterized by intense bursts of high-energy gamma rays of approximately several seconds in duration, they were discovered in the late 1960s by the Vela satellite designed to monitor violations of a nuclear weapons test ban treaty.

Like most observed astrophysical phenomena, empirical classification preceded theoretical understanding. For around 25 years postdiscovery, their origin remained uncertain. Observationally, GRBs are classified into *short* and *long* bursts, according to whether their duration exceeds 2 s, with the majority falling into the latter category. Major advances occurred in the 1990s with the discovery of an isotropic sky distribution implying an extragalactic origin, a finding confirmed in 1997 by the detection of an X-ray and optical *afterglow* in a distant host Galaxy (van Paradijs et al. 1997). With the distance scale established, the implied high luminosities constrained their physical nature. The current understanding is that the short bursts arise from the merger of a pair of neutron stars, while long bursts result from the core collapse of massive stars. Hereafter, we focus on the latter, which are considered to present the greater danger (e.g., Piran and Jimenez 2014).

We have seen that the risk to Earth on timescales of human interest from a supernova is below the current long-term statistical risk of ~1 per Gyr within 10 pc. No such assurances can be given for the GRBs, whose "kill zone" extends much further into the Galaxy. Soon after the realization that GRBs are cosmological, Thorsett (1995) calculated the damaging impact of a Galactic GRB on the Earth's atmosphere. Further modeling of the degree of ozone depletion was presented by Thomas et al. (2005). Piran and Jimenez (2014) used the latter results to classify bursts with integrated gamma-ray fluences (the product of the observed the burst flux, e.g., in W m^{-2}, and its duration in seconds) of 10, 100, and 1000 kJ m^{-2} as capable of causing minor, significant, and catastrophic damage to the biosphere, respectively; they assign 100 kJ m^{-2} as the "canonical life-threatening fluence." Combined with a recent determination of the GRB luminosity function, which specifies the GRB rate per unit cosmological volume as a function of gamma-ray luminosity, they estimated the frequency of terrestrially damaging GRBs within the Milky Way. They reduced the resulting GRB rate within the Milky Way volume by a factor of 10 to account for the fact that the heavy element content (or *metallicity*) of its stellar population overlaps with that of observed GRB host galaxies only at the 10% level; they calculate a 50% probability that a "life threatening" 100 kJ m^{-2} GRB has hit the Earth in the past 500 Myr, a figure that rises to over 90% for the past 5 Gyr (the approximate age of the Earth).

This issue of *metallicity bias*—the hypothesis that long GRBs occur preferentially or perhaps exclusively in low metallicity gas—has been one of the key topics in GRB research over the past decade. The fact that the GRB rate does not precisely track the cosmic star formation rate suggests that they are biased toward a particular type of Galaxy, for example, low mass, low metallicity dwarf galaxies, although the exact dependence remains controversial. In order to gain a more detailed spatiotemporal breakdown of the rate of life-threatening Galactic GRBs, it is necessary to track the evolution of the star formation rate and gas metallicity in the Milky Way's disk. Simulations by Naab and Ostriker (2006) showed that the Milky Way formed from the inside outwards, with high central star formation rates and steep radial gradients at the earliest times, flattening as the Galaxy evolved. Similarly, the radial metallicity gradient is initially steep, but gradually flattens toward its current level. Gowanlock (2016) incorporated these features into a GRB simulation of the Milky Way featuring a 3-D stellar density distribution. In the case where the GRB rate is totally suppressed at high metallicities, he found that there is only a 10% chance that the Earth has been hit by one or more lethal GRBs within the last Gyr, rising to 70% over the last 5 Gyr. At recent times, GRBs are concentrated at large radius due to the metallicity bias.

To assess further the evolving GRB threat facing the Earth, we performed similar calculations to Gowanlock (2016), incorporating the Naab and Ostriker (2006) star formation and metallicity histories. We instead used the results of Graham and Schady (2016) to link the star formation rate to the absolute rate of long-duration GRBs, where the GRB rate drops by a factor of 30 at metallicities greater than half that of the Sun. Whereas Gowanlock (2016) assumed a fixed sterilization distance of 2 kpc at which a typical GRB yields a fluence of 100 kJ m^{-2}, we utilize the full luminosity function for long GRBs as in Piran and Jimenez (2014), implying that sterilizations can occur from GRBs over a much greater extent of the Galaxy. Due to the adopted metallicity bias, the Galactic GRB rate again increases beyond a critical Galactic radius, corresponding to the point where the metallicity drops below threshold. This radius is now 13 kpc, but it was around 8.5 kpc when the Earth formed some 4.6 Gyr ago implying that we were then much closer to life-threatening bursts. The results of our calculations are shown in Table 4.1. As found by Gowanlock (2016), they also suggest that the GRB threat to Earth is currently lower than that estimated by Piran and Jimenez at less than one event every few Gyr, but was substantially higher in the past, especially around the time of Earth's formation. The current rate at which the less-damaging 10 kJ m^{-2} events hit the Earth is around 10 times higher.

So far our discussion has focused entirely on the effects of the gamma rays, but it is highly likely that a Galactic GRB incident upon the Earth would be accompanied by a burst of cosmic rays, which are in turn converted to a burst of ground-level muons. In the context of the "cannonball" model for GRBs, Dar (2008) argues that the cosmic ray risk is actually much

TABLE 4.1 Rate of Life-Threatening (Fluence > 100 kJ m^{-2}) Long-Duration GRBs (in Events per Gyr) for a Planet at Various Galactic Radii as a Function of Cosmic Time

Time before Present (Gyr)	Galactic Radius:3 kpc	5 kpc	8.5 kpc	13 kpc
0	0.18	0.18	0.18	0.32
1.6	0.26	0.27	0.34	0.59
3.6	0.42	0.46	0.74	0.68
5.6	0.74	0.92	1.36	0.59

Note: The Sun lies at 8.5 kpc.

more severe, by several orders of magnitude in terms of fluence, than that posed by the gamma rays. As a result, the above calculations may well substantially underestimate the danger posed by GRBs. Traveling at almost the speed of light the cosmic rays from a Galactic burst would be delayed, perhaps by as little as a few days, relative to the gamma rays. A situation may arise in which an initial flash of gamma rays, which alone are not sufficient to cause major damage, presage the imminent arrival of a catastrophic dose of cosmic rays.

4.2.3 Stirrings of the Monster: Outbursts from the Galactic Center Black Hole and Galaxy-Wide Feedback Events

Supernovae and gamma-ray bursts typically leave behind black holes with masses a few times that of the Sun, or in extreme cases a few tens of Solar masses. When fed by gas, for example by mass transfer from a companion star, they emit high-energy radiation as the accreted material is compressed and heated as it spirals in toward the black hole. Such *stellar mass black holes* are unlikely to present a danger to Earth, except in the rare case where the accretion in a nearby system leads to the formation of relativistically beamed jets pointing toward us, analogous to those of a GRB. Such sources include *microquasars*, and some *ultraluminous X-ray sources;* for a review see Mirabel and Rodriguez (1999). Other high-energy transient phenomena, whose nature is not fully understood but thought to involve a stellar remnant (neutron star or black hole) include ultraluminous X-ray bursts (Irwin et al. 2016) and fast radio bursts (Lorimer et al. 2007).

At the other end of the scale are *supermassive* black holes, some 10^6–10^{10} times as massive as the Sun. They reside in the nuclei of galaxies and their masses correlate with the properties of their host galaxies, for reasons probably related to so-called "feedback" processes during their major growth phase at earlier cosmic epochs, when they accreted gas and shone as *Active Galactic Nuclei* (AGN) (Fabian 2012). At present, most Galactic nuclei are dormant, including the 4 million Solar mass black hole at the center of own Galaxy. This source, known as Sgr A* due to its location in the constellation Sagittarius, has been intensively monitored at radio and X-ray wavelengths in recent decades, and the mass of this black hole inferred by tracking the orbits of stars in its neighborhood at near-infrared wavelengths (Ghez et al. 2008; Gillessen et al. 2009).

The present radiative output of Sgr A* is $\sim$9 orders of magnitude below the so-called Eddington limit, the level which characterizes a well-fed radiatively efficient AGN*; nevertheless, its luminosity can increase by a factor of ~100 during flaring episodes lasting several hours—see, for example, the review of Sgr A* activity by Ponti et al. (2013). There is evidence, however, that Sgr A* has been much more luminous in the past. The first such evidence comes in the form of X-ray reflection signatures from the dense gas in the inner few hundred parsecs of the Galaxy, the so-called *central molecular zone* (*CMZ*). The CMZ obscures the sightline to Sgr A* from most vantage points in the Galactic plane, rendering it completely invisible at optical or even X-ray wavelengths. In contrast, the obscuration along our current sightline is lower, permitting study, although optical emission is still heavily suppressed. The CMZ can act as a mirror reflecting X-rays from Sgr A* with a characteristic signature, which reach us with a time delay due to the extra light travel time relative to emission directly from Sgr A*. Detailed X-ray studies of these reflection features (e.g., Terrier et al. 2010) show that Sgr A* was about a million times more luminous approximately 100 years ago. Although it would have been one of the brightest persistent X-ray sources in the sky at the time, its X-ray flux was still ~1 million times fainter than that of the Sun (at Solar Minimum†) and hence of no danger to the Earth.

Moving to more ancient epochs, there is evidence that Sgr A* shone even more intensely as a fully fledged AGN or Seyfert Galaxy some 6 million years ago in a phase lasting several million years. The first comes from the discovery in 2010 by the Fermi gamma-ray telescope of a pair of giant bubbles extending perpendicularly to the Galactic plane and originating in its nucleus (Su, Slatyer, Finkbeiner 2010), due to the input of cosmic rays from the AGN and/or a coeval burst of nuclear star formation. Another indirect indicator of past nuclear activity comes from the detection of emission from a broad arch of gas in the so-called Magellanic Stream, whose intensity is consistent with illumination by an active Sgr A* (Bland-Hawthorn et al. 2013); finally, the distribution of million degree gas around the Milky Way is consistent with Sgr A* having blown a cavity of ~6 kpc radius in this low density material during its outburst (Nicastro et al. 2016). Emitting at close to the Eddington limit, the X-ray flux at Earth from Sgr A* would have rendered it the brightest X-ray source in the sky, and just 1–2 orders of magnitude fainter than the Sun (see also Amaro-Seoane

* The Eddington Limit is the luminosity at which outward radiation pressure balances the inward force of gravity due to the black hole.

† See the compilation of the X-ray fluxes for the brightest observed sources at https://heasarc.gsfc.nasa.gov/docs/heasarc/htheadates/brightest.html.

and Chen 2014). The outburst may have been triggered by a molecular cloud from the CMZ falling into the inner parsec of the Galaxy (Zubovas and Nayakshin 2012).

These recurrent outbursts from Sgr A* have not posed a hazard to Earth unless accompanied by relativistic jets directed toward us, as portrayed by Fred and Geoffrey Hoyle in their science fiction work, *The Inferno* (Hoyle and Hoyle 1973). Looking to the far future, recent simulations of Galaxy formation and black hole growth suggest that Sgr A* could enter a major growth phase which would have dramatic consequences for the entire Galaxy (Bower et al. 2017). The latter model seeks to explain the observed dichotomy between "blue" galaxies, with copious ongoing star formation, and "red" galaxies, in which star formation has all but ceased, by invoking feedback from black hole accretion to sweep the host Galaxy clear of gas. The key parameter is the mass of the Galaxy's *dark matter halo,* which accounts for around 80% of its entire mass. Below a critical halo mass of around 10^{12} Solar masses, the rate at which cool gas enters the Galaxy is mostly balanced by a buoyant outflow of gas driven by the star formation. As a result, the gas density around the black hole cannot build up to fuel a major outburst. Above the critical dark halo mass the outflow is no longer buoyant, trapping gas in the inner Galaxy and triggering a phase of rapid nonlinear black hole mass growth. In practice the black hole cannot grow indefinitely, and the outburst terminates once its feedback, via jets and winds, has driven out the bulk of the cool gas from the Galaxy and largely shut down star formation.

Intriguingly, the halo mass of the Milky Way is close to the critical 10^{12} Solar mass transition threshold (e.g., Carlesi et al. 2016), above which its 4 million Solar mass black hole would grow in mass by a factor of 10–100 within ~400 Myr. Emitting at around the Eddington limit as it grows, it would, by the end of this time have an observed X-ray flux comparable with that of the present-day Sun, and potentially much higher during flares (AGN are highly variable sources, due to fluctuations in accretion rate and associated disk instabilities). A possible trigger for this event may be the collision or merger between the Milky Way and Andromeda Galaxies, which is expected to occur within a few billion years prior to the end of the Sun's main sequence life. According to simulations by Cox and Loeb (2008), this collision may result in the Sun being expelled to a much larger radius (>30 kpc) within the resulting elliptical Galaxy.

Another form of Galaxy-wide feedback occurs in *starburst galaxies.* These are galaxies undergoing bursts of locally intense star formation, usually in their nuclear regions and often triggered by interaction or merger with a companion Galaxy. When the local intensity of star formation per unit area of Galactic disk exceeds a critical threshold, the energy input from stellar winds and supernovae can heat the interstellar medium (ISM) in the starburst region to temperatures of hundreds of millions K. As a result, this superheated metal-rich gas becomes overpressurized relative to the ambient ISM and flows out of the Galaxy via the path of least resistance, usually perpendicular to the disk as a bipolar outflow. Observations reveal multiphase gas outflowing at speeds of several hundred km/s, dominated by the cool ISM material swept up in the flow. For a recent review see Heckman and Thompson (2017).

A critical issue is whether the superwinds have sufficient speed to escape their host Galaxy and spread heavy elements within intergalactic space. If they do not, the outflowing gas ultimately falls back into the Galaxy and enriches the ISM for the next generation of stars. This gives rise to an observed Galaxy mass–metallicity relation: metals readily escape the low mass haloes (e.g., dwarf galaxies) keeping them metal poor, but are retained within more massive galaxies. As a result, the Milky Way lies at the upper end of a relatively narrow band ($\sim$2 orders of magnitude wide) in Galaxy stellar mass where the dangers from GRB and AGN feedback are both relatively low. In galaxies $\geq$100 times less massive, the low metallicity leads to an elevated GRB risk; galaxies more massive than the Milky Way have typically undergone a phase of nonlinear black hole growth, shutting off star and planet formation.

4.2.4 Interaction with the Galactic Environment: Dynamical Perturbations and Variations in the Local Interstellar Medium

So far we have focused on phenomena which may be termed *radiative transients*, ranging from explosions such as supernovae and GRBs where the hazard evolves significantly on timescales of human interest, through to longer-lived ($>10^6$ years but still *cosmologically* transient) outbursts from supermassive black hole accretion and starburst superwinds. We now briefly review the risks to the Earth from: (1) dynamical perturbations to the Solar System from the Galaxy and its constituents, which may push comets onto Earth-impacting orbits; and (2) slow secular variations, on timescales of tens to hundreds of Myr, in the local interstellar medium (e.g., background cosmic ray intensity) as the Sun orbits the Galaxy.

The Milky Way Galaxy as a whole is a *collisionless stellar system*, where two-body interactions that significantly perturb the motion of the two stars are unimportant. Nevertheless, given the extent of the Solar System's outer reaches—the Oort Cloud cometary reservoir extends from tens to hundreds of thousand AU, more than 10% of the distance to the nearest star—even close "stellar flybys" rather than actual collisions can have significant effects. Indeed, Mamajek et al. (2015) discovered that some 70,000 years ago a low mass binary system passed within 0.25 pc of the Sun, likely within the outer reaches of the Oort Cloud. Although they claim that this event probably had a negligible impact on the long-term cometary flux, they note that the expected rate of such close encounters is once per 10 Myr, so this was a fortuitously recent occurrence.

Quite apart from external perturbations, including those from massive bodies in the outer reaches of the Solar System such as the recently hypothesized 10 Earth-mass "Planet 9" (Batygin and Brown 2016), the known planets of the Solar System constitute an intrinsically chaotic system. This limits the accuracy of numerical predictions of past and future planetary motions beyond a few tens of Myr. The Solar System may also be unstable, and planetary collisions or ejections may occur, as reviewed by Laskar (2012) who estimates that such catastrophic outcomes have an occurrence probability of about 1% over the next 5 Gyr.

As the Sun orbits the Galaxy with a period of ~225 Myr, it oscillates above and below the mid-plane of the Galactic disk with a period of ~70 Myr. Numerous authors have sought to correlate this motion and the resulting variations in the Galactic tidal gravitational field and the stellar and molecular cloud encounter rate, with claimed 25–35 Myr periodicities in impact-triggered terrestrial mass extinctions; see, for example, Wickramasinghe and Napier (2008) and Rampino (2015) and references therein. Such studies suggest that we are currently close to the plane and near the peak of a bombardment episode. It has also been claimed (e.g., Randall and Reece 2014) that some of the Milky Way's dark matter comprises a very thin (~10 pc) dissipative disk which enhances the plane-crossing signal; Rampino (2015) has proposed that capture by the Earth's core of particularly dense clumps of dark matter during such plane-crossings may lead to enhanced internal heating, driving geological activity (e.g., volcanism) with a similar periodicity to that of the mass extinctions.

In addition to vertical oscillation through the Galactic plane, the Sun also passes in and out of the Galaxy's spiral arms during its orbit. With the bulk of the star formation and supernovae occurring in the arms, this leads to an increase in the cosmic ray flux, which may trigger cloud formation and drive terrestrial climate variations. Numerous studies have sought to associate variations in past climate indicators and meteoritic cosmogenic isotope abundances with the expected time interval of ∼135 Myr between spiral arm crossing and global variations in the Milky Way's past star formation rate (see e.g., Shaviv 2002; Scherer et al. 2006). Within the spiral arms, there is also a higher probability of the Solar System encountering a dense molecular cloud; the deposition of dust in the atmosphere may trigger climatic cooling and possibly a transition to a snowball state (e.g., Pavlov et al. 2005).

4.3 THE HAZARD-HABITABILITY NEXUS

4.3.1 The Galactic Habitable Zone

Until we have detected primitive or intelligent life elsewhere in the Galaxy, efforts to understand its distribution must focus on finding and characterizing those habitats where it may be able to exist. As such, the concept of *habitability* is a key driver of much exoplanetary and astrobiological research. At the level of the individual star-exoplanet system, the concept of the *Circumstellar Habitable Zone* (*CHZ*), characterizing the range of orbital radii around the star within which liquid water at a pressure of 1 bar could exist on a planet, is well developed; it is a function of stellar and orbital properties which can be readily measured. It is of course a necessary but not sufficient condition for the presence of life, and in practice any number of additional currently unobservable factors may render a planet in the CHZ inhospitable, including but not limited to atmospheric composition, orbital instability, and plate tectonics. Indeed, Ward and Browlee (2000) argued that while primitive life may be widespread, the prerequisites for the development of complex life may be so exacting as to make it exceedingly rare in the Galaxy (see also simulations by Forgan and Rice 2010).

By analogy with the CHZ, the concept of a *Galactic Habitable Zone* (*GHZ*) has gained traction, beginning with the work by Gonzalez et al. (2001). The GHZ demarcates the region of the Galaxy containing suitable host stars (similar in spectral type to the Sun) of sufficient age (∼4 Gyr) to allow for the development of complex life, sufficient metallicity for terrrestrial planet formation (but not so high that close-in giant planets pose a threat to Earth-mass planets), and with a sufficiently low rate of hazardous nearby supernovae. The simulations of Lineweaver, Fenner, and Gibson (2004) showed that such criteria define a GHZ in the form of an annular zone between 7 and 9 kpc from the Galactic Center, and that the bulk of the habitable stars in this zone are older than the Sun by ∼1 Gyr.

Others have cast doubt on the existence of such a well-defined GHZ. Monte Carlo simulations by Gowanlock et al. (2011), taking into account a 3-D Galactic structure, did not find a sharp inner boundary, with the bulk of the habitable planets in fact residing in the inner Galaxy (see also Prantzos 2008). Using high-resolution numerical simulations of the Milky Way and the nearby spiral Galaxy M33, Forgan et al. (2017) showed that their GHZs lack azimuthal symmetry and depend on the galaxies' detailed evolutionary histories; habitable planets can be found in the inner regions and in structures such as spiral arms, tidal streams, and satellite galaxies. While the GHZ may be more complicated than originally assumed, the concept has served to highlight the factors influencing habitability on such scales; future simulations need to take account of the local and potentially global impact of AGN outbursts and GRBs, not just the local effects of supernovae. Nevertheless, the existing work shows that the principal ingredients and conditions for complex life in the Milky Way are relatively widespread, and that most such life could well be several billion years more evolved than us. This has implications for the search for intelligent life elsewhere in the Galaxy, as discussed in Section 4.4.

4.3.2 The Habitability of the Universe through Cosmic Time

As a natural extension of these Galactic studies, several authors have begun to assess how the habitability of the entire Universe evolves over time. Dayal et al. (2016) considered the cosmic evolution of the principal hazards (supernovae, gamma-ray bursts, and AGN), as illustrated in Figure 4.2. As seen from this figure, supernovae are the most frequent hazard at any cosmic epoch and the overall hazard rate has been in steady decline for the last 10 billion years. On the other hand, the stellar mass density, which is a proxy for the total number of stars around which planets will form, continually grows through time. Taken together, this results in a scenario where the habitability of the Universe has been continually rising through cosmic time, and will continue to do so for several hundred billion years until the final generation of stars dies out.

The review of cosmic habitability by Loeb (2016) focused instead on the cosmological evolution of the key ingredients for life, rather than the hazards. He concluded that the Universe will become more habitable in the future, provided that life can take hold around the abundant population of low-mass stars with the longest lifetimes. Only if life is unfeasible around such stars (whose masses extend down to as little as 8% of the mass of the Sun) would we be living in the most likely epoch for life. While observations

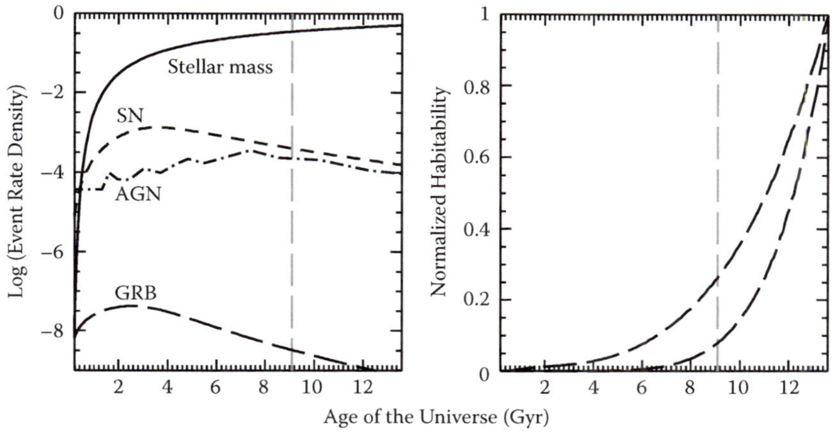

FIGURE 4.2 Left panel shows the evolution with cosmic time of the number densities of supernovae and GRBs (in units of events per year per Mpc^3), and AGN (number of sources per Mpc^3). The solid line shows the total stellar mass density (in units of 10^9 solar mass per Mpc^3). Right panel shows the probability of habitability of the Universe at any epoch, normalized to the present day, for reasonable habitability scenarios. The vertical dashed line shows the age of the Universe (9.1 billion yr) when the Earth formed.

suggest that rocky planets are abundant around low-mass (M dwarf) stars, their habitability is under debate due to the impact of, for example, stellar flaring and tidally locked planetary rotation (Shields et al. 2016).

4.4 RISKS FROM INTELLIGENT LIFE ELSEWHERE IN THE GALAXY

The search for intelligent life elsewhere in the Universe is arguably the chief underlying motive for much current exoplanetary research. Therefore, having reviewed the purely astrophysical hazards facing the Earth, we complete our coverage of Galactic risks by considering those stemming from the actions of intelligent life elsewhere in the Galaxy.

While the potential existence of extraterrestrial life has long been a matter of speculation among philosophers, theologians, and science fiction enthusiasts, the prospects for its discovery, and their implications, are now taken more seriously than ever. As reviewed by Shostak (2015), the initial discovery is likely to come through one of three channels: (1) the in situ detection of primitive life within the Solar System; (2) the spectroscopic detection of biomarkers in exoplanet atmospheres; and (3) the direct detection of intelligent life either via their intentional electromagnetic signals, as sought by SETI programs, or by discovering evidence for feats of intelligent "astroengineering" visible over interstellar or intergalactic distances, which defy our understanding of natural astrophysical phenomena. The first of these discovery channels may pose challenges with regard to the issues of cross-contamination and planetary protection, including the possibility of back-contamination to Earth from a sample return mission (see Race 2015 for further discussion); the second poses no immediate risk in and of itself, while the third may give rise to all manner of consequences. We first consider what, if anything, we can already infer from its apparent absence, before reviewing the risks which may stem from its existence and our discovery of it.

4.4.1 The Drake Equation and the Fermi Paradox: Rationalizing the Silence

SETI programs have been operating intermittently for over half a century now, thus far with no positive detections (for a review of the first four decades of search see Tarter 2001). The optimism of the early years inspired pioneers such as Frank Drake to attempt to estimate the number of communicating civilizations in our Galaxy, culminating in the celebrated Drake Equation. The latter gives the number of communicating civilizations as $N_{comm} = R^{*}fL$, where R^{*} is the rate at which stars are forming in the Galaxy, and f is the product of the number of habitable planets per star and the fraction with intelligent, technologically advanced, communicating civilizations; L is the lifetime of the communication phase. Of these terms, only R^{*} is known to any degree of confidence while the crucial f factor ranges anywhere from 0 to $\sim$1, consistent with an enormous range of possible values for N_{comm}. Although sometimes considered as little more than a parameterization of our ignorance the Drake Equation remains a useful organizational framework, and with recent advances in exoplanet characterization some of the factors that contribute to the f term are becoming less speculative (Vakoch and Dowd 2015).

The tension between the apparent lack of any trace of evidence for extraterrestrial intelligence and the short time required to colonize the Milky Way Galaxy ($\sim$millions of years, at a modest fraction of the speed of light) is known as the Fermi Paradox. It is often discussed in the same context as the Drake Equation, and there is no shortage of explanations, spanning the full range of astrophysical, astrobiological, sociological, and technological considerations. Recent discoveries of the abundance of exoplanets, most of which may be several billion years older than Earth, have rendered the paradox more acute. One astrophysical explanation is that the Galactic GRB rate has so far been high enough to repeatedly sterilize large parts of the Galaxy before life could reach the necessary levels of complexity; if intelligent life requires ~1 Gyr to evolve and with the Galactic GRB rate in decline (as discussed in Section 4.2.2), it has been suggested that we may be on the cusp of a phase transition toward a state in which the Galaxy could be readily colonized (e.g., Annis 1999; Ćirković 2008; and references therein).

On the other hand, if intelligent life has been able to develop and flourish in spite of such hazards, it is likely to be highly advanced, given that most of the habitable stars in the Galaxy are at least 1 Gyr older than the Sun (see Section 4.3.1). As such, the silence may result from our civilization representing a comparatively brief transition between primitive microbial life and some form of highly sophisticated superintelligence (see for example, Vidal 2015); the latter may have no interest in communicating with us and may also be difficult to distinguish from natural astrophysical phenomena, as for example with Hoyle's fictional *Black Cloud* (Hoyle 1957). While there is no current evidence for astroengineering artefacts on stellar or Galactic scales, the opening of new regimes of astronomical survey parameter space and targeted observations of exoplanets may bring such evidence to light. A recent case has centered on the star KIC 8462852 (known as Boyajian's star) discovered by the *Kepler* planetary transit mission, which exhibits irregular large-amplitude dimming events (Boyajian et al. 2016); one possibility is that these light dips are due to obscuration by artificial circumstellar structures, although variations in the intervening interstellar medium may offer a more plausible explanation (Wright and Sigurdsson 2016). Lingam and Loeb (2017) proposed that the fast radio bursts referred to in Section 4.2.3 may be artificial in origin. Beech (2011) has suggested that searches for artificial structures could be directed toward stars within 10–30 pc of presupernova candidates (or young supernova remnants) as any civilizations there seek protection from their hazardous effects.

At the current time, just two decades after the first exoplanet discoveries and half a century after the start of SETI, with many possible explanations for the absence of signs of extraterrestrial intelligence, it is premature to draw firm conclusions. The curiosity and the motivation to search are as strong as ever, at least among some in our society, as exemplified by the Breakthrough Initiatives previously mentioned. Yet, it is possible to imagine that after several more decades of intensive but ultimately fruitless searching the silence may become palpable, especially if by then the biomarkers of primitive life appear to be a common feature of exoplanet atmospheres. The latter observations may address deeper questions about the origin of life on Galactic scales: it has been shown that the observed degree of spatial clustering of stars hosting planets with biomarkers could be used to assess whether life arose independently in each stellar system, or whether it spread between them from a smaller number of seed sites as natural life propagation ("panspermia") or technological colonization (see e.g., Lin and Loeb 2015; Lingam 2016).

If intelligent life does exist elsewhere in the Galaxy, what risks could it pose to us and what are the risks in searching for it? Some of the former risks have been addressed by Ćirković (2008); he plays down the possibility (or even the feasibility) of interstellar conflict, but suggests that we should remain alert to the threats posed by "deadly probes"—fleets of self-replicating spacecraft launched by an intelligent agency, which by accident or design could destroy other civilizations. Probes in general, whether hostile or benign, warrant further attention and observational search effort within the Solar System. This has been advocated by Gertz (2016a) who argues that they represent a more efficient means for information exchange than interstellar signaling and conventional SETI. However, Kowald (2016) has suggested that runaway error propagation in any self-replicating system with finite accuracy would ultimately limit their ability to colonize.

SETI programs pose no intrinsic risk until an intelligent signal is discovered. At that point, the calculus would change dramatically. A framework of policies and protocols which would be triggered upon discovery is in place, developed under the auspices of the International Academy of Astronautics (IAA) and requiring notification of the relevant United Nations (UN) bodies. It is a code of self-governance; however, none of it has been officially incorporated into the UN framework and none of it is in any sense legally binding. For an excellent review of this framework and suggestions for improving it, see Race (2015). Active SETI, or messaging to extraterrestrial intelligence (METI), is somewhat more controversial; the community is divided over the risks and ethics of this activity and this is an area where SETI's mutually agreed code of self-governance may be breaking down (Race 2015; Gertz 2016b,c).

4.5 CONCLUSIONS

We have seen that according to current models the threat to Earth posed by hazards such as supernovae and gamma-ray bursts is minimal; existentially threatening events of this nature are expected to impact the Earth around once per Gyr at the current epoch, and roughly twice as frequently when the Earth was forming around 5 Gyr ago due to the increased gamma-ray burst rate. The supermassive black hole at the Galactic Center, Sgr A*, also poses no threat in its current intermittently fed state, but it may at some future point undergo a phase transition in accretion rate during which it would grow exponentially to 10 to 100 times its current mass in just a few hundred million years. During this phase its X-ray brightness would rival the Sun's and would be much higher during flares or if accompanied by jets directed at Earth. A trigger for this episode may be the collision between our Milky Way and the neighboring Andromeda Galaxy in a few billion years time.

Although much of the above is at first reassuring for the future of our civilization, we end with two brief caveats. The first concerns *anthropic bias*. As discussed by Bostrom and Ćirković (2008) and Ćirković (2008), this occurs when we ignore fundamental selection effects—for example, the fact that we have evolved for long enough to be able to consider these matters cannot in and of itself be used to make any general inferences concerning the rarity of existential Galactic hazards—we may just be an extremely lucky outlier, while life on most Earth-like planets elsewhere in the Galaxy has been destroyed at an earlier stage of development. Of course, astrophysical understanding can be used to strengthen claims stemming from our own apparent longevity, but such arguments must be balanced against the completeness of this knowledge.

The second caveat thus concerns the completeness of our astrophysical understanding and the potential for new and unexpected discoveries. It is only just over half a century since X-ray and radio astronomy opened the window on the violent high-energy Universe, bringing to light unanticipated exotica such as accreting black holes and gamma-ray bursts. It is possible that as other new windows on the Universe are opened, hazardous phenomena may come to light, either naturalistic in origin or due to the actions of intelligent agencies, and we should not be biased against explanations of the latter kind a priori.

For example, the development of gravitational wave astronomy will probe the structure of space time and may reveal phenomena entirely inaccessible to electromagnetic observation. Conventional electromagnetic astronomy is also poised to probe new regimes of observational parameter space, not just in terms of sensitivity and resolution, but in the *time domain* thanks to the development of facilities such as the Square Kilometer Array (SKA)* radio telescope and the Large Synoptic Survey Telescope (LSST)† at optical wavelengths. By surveying

* https://www.skatelescope.org/.

† https://www.lsst.org/.

large areas of the sky on short timescales such facilities will probe the time variability of astrophysical sources, with the potential for qualitatively new discoveries. As argued by Norris (2017), instrumental complexity and large data volumes will, however, make such discoveries unlikely to happen by chance during the pursuit of planned science programs; instead, he argues that explicit methods for uncovering new sources and new phenomena (including those related to SETI) need to be incorporated into the next-generation astronomical surveys (e.g., using machine learning techniques) if these facilities are to reach their full potential.

ACKNOWLEDGMENT

PD acknowledges support from the European Research Council's starting grant ERC StG-717001 "DELPHI" and from the European Commission's and University of Groningen's CO-FUND Rosalind Franklin program.

REFERENCES

Abbott, B.P., Abbott, R., Abbott, T.D. et al. 2016. Observation of gravitational waves from a binary black hole merger. *Phys. Rev. Lett.* 116, 061102.

Adams, S.M., Kochanek, C.S., Beacom, J.F., Vagins, M.R., Stanek, K.Z. 2013. Observing the next galactic supernova. *Astrophys. J.* 778, 164–178.

Amaro-Seoane, P. and Chen, X. 2014. *Our Supermassive Black Hole Rivaled the Sun in the Ancient X-ray Sky.* https://arxiv.org/abs/1412.5592 (accessed November 7, 2017).

Annis, J. 1999. An astrophysical explanation for the great silence. *J. Brit. Interplan. Soc.* 52, 19–22.

Batygin, K. and Brown, M.E. 2016. Evidence for a distant giant planet in the solar system. *Astron J.* 151, 2, 22.

Beech, M. 2011. The past, present and future supernova threat to earth's biosphere. *Astrophys. Space Sci.* 336, 287–302.

Bland-Hawthorn, J., Maloney, P.R., Sutherland, R.S., Madsen, G.J. 2013. Fossil imprint of a powerful flare at the galactic center along the magellanic stream. *Astrophys. J.* 778, 58.

Bonnet, R.-M. and Woltjer, L. 2008. *Surviving 1000 Centuries Can We Do It?* Chichester, UK: Praxis Publising, Springer.

Bostrom, N. and Ćirković, M.M. 2008. Introduction. In *Global Catastrophic Risks*, eds. N. Bostrom and M.M. Ćirković, Oxford: Oxford University Press, pp. 1–29.

Bower, R.G., Schaye, J., Frenk, C.S. et al. 2017. The dark nemesis of galaxy formation: Why hot haloes trigger black hole growth and bring star formation to an end. *Mon. Not R Astron. Soc.* 465, 1, 32–44.

Boyajian, T.S., LaCourse, D.M., Rappaport, S.A. et al. 2016. Planet Hunters IX. KIC 8462852- where's the flux? *Mon. Not R Astron. Soc.* 457, 3988–4004.

Breitschwerdt, D., Feige, J., Schulreich, M.M., de Avillez, M.A., Dettbarn, C., Fuchs, B. 2016. The locations of recent supernovae near the sun from modelling 60Fe transport. *Nature* 532, 73–76.
Carlesi, E., Hoffman, Y., Sorce, J.G., Gottlöber, S. 2016. Constraining the mass of the local group. *Mon. Not R Astron. Soc.* 465, 4886–4894.
Ćirković, M.M. 2008. Observation selection effects and global catastrophic risks. In *Global Catastrophic Risks*, eds. N. Bostrom and M.M. Ćirković, Oxford: Oxford University Press, pp. 120–145.
Cox, T.J. and Loeb, A. 2008. The collision between the Milky Way and andromeda. *Mon. Not R Astron. Soc.* 386, 1, 461–474.
Crawford, I.A. 2017. The moon as a recorder of nearby supernovae. In *Handbook of Supernovae*, eds. A.W. Alsabti and P. Murdin. Cham, Switzerland: Springer International Publisher, pp. 2507–2522.
Dar, A. 2008. Influence of supernovae, gamma-ray bursts, solar flares, and cosmic rays on the terrestrial environment. In *Global Catastrophic Risks*, eds. N. Bostrom and M.M. Ćirković, Oxford: Oxford University Press, pp. 238–261.
Dayal, P., Ward, M.J., Cockell, C. 2016. *The Habitability of the Universe through 13 Billion Years of Cosmic Time*. https://arxiv.org/abs/1606.09224.
Fabian, A.C. 2012. Observational evidence of active galactic nucleus feedback. *Annu. Rev. Astron. Astrophys.* 50, 455–489.
Fields, B.D., Athanassiadou, T., Johnson, S.R. 2008. Supernova collisions with the heliosphere. *Astrophys. J.* 678, 549–562.
Forgan, D.H. and Rice, K. 2010. Numerical testing of the rare earth hypothesis using monte carlo realization techniques. *Int. J. Astrobiol.* 9, 73–80.
Forgan, D.H., Dayal, P., Cockell, C., Libeskind, N. 2017. Evaluating galactic habitability using high-resolution cosmological simulations of galaxy formation. *Int. J. Astrobiol.* 16, 60–73.
Fry, B.J., Fields, B.D., Ellis, J.R. 2015. Astrophysical shrapnel: Discriminating among near-earth stellar explosion sources of live radioactive isotopes. *Astrophys. J.* 800, 71–87.
Fry, B.J., Fields, B.D., Ellis, J.R. 2016. Radioactive iron rain: Transporting 60Fe in supernova dust to the ocean floor. *Astrophys. J.* 827, 48–64.
Gehrels, N., Laird, C.M., Jackman, C.H., Cannizzo, J.K., Mattson, B.J., Chen, W. 2003. Ozone depletion from nearby supernovae. *Astrophys. J.* 585, 1169–1176.
Gertz, J. 2016a. ET probes: Looking here as well as there. *J. Brit. Interplan. Soc.* 69, 88–91.
Gertz, J. 2016b. Reviewing METI: A critical analysis of the arguments. *J. Brit. Interplan. Soc.* 69, 31–36.
Gertz, J. 2016c. Post-detection SETI protocols & METI: The time has come to regulate them both. *J. Brit. Interplan. Soc.* 69, 263–270.
Ghez, A.M., Salim, S., Weinberg, N.N. et al. 2008. Measuring distance and properties of the Milky Way's central supermassive black hole with stellar orbits. *Astrophys. J.* 689, 1044–1062.
Gillessen, S., Eisenhauer, F., Trippe, S. et al. 2009. Monitoring stellar orbits around the massive black hole in the Galactic Center. *Astrophys. J.* 692, 1075–1109.

Gonzalez, G., Brownless, D., Ward, P. 2001. The Galactic Habitable Zone: Galactic chemical evolution. *Icarus* 152, 185–200.

Gowanlock, M.G. 2016. Astrobiological effects of gamma-ray bursts in the Milky Way galaxy. *Astrophys. J.* 832, 38.

Gowanlock, M.G., Patton, D.R., McConnell, S.M. 2011. A model of habitability within the Milky Way Galaxy. *Astrobiol.* 11, 855–873.

Graham, J.F. and Schady, P. 2016. The absolute rate of LGRB formation. *Astrophys. J.* 823, 154.

Heckman, T.M. and Thompson, T.A. 2017. Galactic winds and the Role Played by Massive Stars. In *Handbook of Supernovae*, eds. A.W. Alsabti and P. Murdin. Cham, Switzerland: Springer International Publishe, pp. 2431–2454.

Hoyle, F. 1957. *The Black Cloud*. William Heinemann.

Hoyle, F. and Hoyle, G. 1973. *The Inferno*. William Heinemann.

Hurley, K., Boggs, S.E., Smith, D.M. et al. 2005. An exceptionally bright flare from SGR 1806–20 and the origins of short-duration γ-ray bursts. *Nature* 434, 1098–1103.

Inan, U., Lehtinen, N., Moore, R. et al. 2005. Massive disturbance of the daytime lower ionosphere by the giant x-ray flare from magnetar SGR 1806–20, *IAGA Toulouse*, France.

Irwin, J.A., Maksym, W.P., Sivakoff, G.R. et al. 2016. Ultraluminous X-ray bursts in two ultracompact companions to nearby elliptical galaxies. *Nature* 538, 7625, 356–358.

Kaltenegger, L. 2017. Characterizing habitable worlds. *Annu. Rev. Astron. Astrophys.* 55, 433–485.

Kaspi, V.M. and Beloborodov, A.M. 2017. Magnetars. *Annu. Rev. Astron. Astrophys.* 55, 261–301.

Kennicutt, R.C. Jr, Evans, N.J. II 2012. Star formation in the Milky Way and nearby galaxies. *Annu. Rev. Astron. Astrophys.* 50, 531–608.

Knie, K., Korschinek, G., Faestermann, T., Wallner, C., Scholten, J., Hillebrandt, W. 1999. Indication for supernova produced 60Fe activity on earth. *Phys. Rev. Lett.* 83, 18–21.

Korschinek, G. 2017. Mass Extinctions and Supernova Explosions. In *Handbook of Supernovae*, eds. A.W. Alsabti and P. Murdin. Cham, Switzerland: Springer International Publisher, pp. 2419–2430.

Kowald, A. 2016. Why is there no von Neumann probe on Ceres? Error catastrophe can explain the Fermi-Hart Paradox. *J. Brit. Interplan. Soc.* 68, 383–388.

Laskar, J. 2012. Text of the Lecture given in the Poincaré Séminar, Paris, on June 2010 (https://arxiv.org/abs/1209.5996).

Lin, H.W. and Loeb, A. 2015. Statistical signatures of panspermia in exoplanet surveys. *Astrophys. J. Lett.* 810, L3.

Lineweaver, C.H., Fenner, Y., Gibson, B.K. 2004. The Galactic Habitable Zone and the age distribution of complex life in the Milky Way. *Science* 303, 59–62.

Lingam, M. 2016. Interstellar travel and galactic colonization: Insights from percolation theory and the yule process. *Astrobiol.* 16, 418–426.

Lingam, M. and Loeb, A. 2017. Fast radio bursts from extragalactic light sails. *Astrophys. J. Lett.* 837, 23.

Loeb, A. 2016. *On the Habitability of Our Universe.* https://arxiv.org/abs/1606.08926.

Lorimer, D.R., Bailes, M., McLaughlin, M.A., Narkevic, D.J., Crawford, F. 2007. A bright millisecond radio burst of extragalactic origin. *Science* 318, 5851, 777–780.

Mamajek, E.E., Barenfeld, S.A., Ivanov, V.D. et al. 2015. The closest known flyby of a star to the solar system. *Astrophys. J. Lett.* 800, L17–20.

Mandea, M. and Balasis, G. 2005. The SGR 1806–20 magnetar signature on the earth's magnetic field. *Geophys J. Int.* 167(2), 586–591.

Melott, A.C., Thomas B.C., Kachelrieß, M., Seminkoz, D.V., Overholt, A.C. 2017. A supernova at 50 pc: Effects on the earth's atmosphere and biota. *Astrophys. J.* 840, 105.

Melott, A.L. and Thomas, B.C. 2011. Astrophysical ionizing radiation and earth: A brief review and census of intermittent intense sources. *Astrobiol.* 11, 343–361.

Mirabel, I.F. and Rodriguez, L.F. 1999. Sources of relativistic jets in the galaxy. *Annu. Rev. Astron. Astrophys.* 37, 409–443.

Naab, T. and Ostriker, J.P. 2006. A simple model of the evolution of disk galaxies: The Milky Way. *Mon. Not R Astron. Soc.* 366, 899–917.

Nicastro, F., Senatore, F., Krongold, Y., Mathur, S., Elvis, M. 2016. A distant echo of Milky Way central activity closes the galaxy's baryon census. *Astrophys. J. Lett.* 828, 12.

Norris, R.P. 2017. Discovering the unexpected in astronomical survey data. *Publ. Astron. Soc. Aus.* 34, 7.

Palmer, D.M., Barthelmy, S., Gehrels, N. et al. 2005. A giant γ-ray flare from the magnetar SGR 1806–20. *Nature* 434, 1107–1109.

Pavlov, A.A., Toon, O.B., Pavlov, A.K., Bally, J., Pollard, D. 2005. Passing through a giant molecular cloud: "Snowball" glaciations produced by interstellar dust. *Geophys. Res. Lett.* 32, L03705.

Piran, T. and Jimenez, R. 2014. Possible role of gamma ray bursts on life extinction in the universe. *Phys. Rev. Lett.* 113, 231102.

Ponti, G., Morris, M.R., Terrier, R., Goldwurm, A. 2013. Traces of Past Activity in the Galactic Centre. In *Cosmic Rays in Star-Forming Environments. Astrophysics and Space Science Proceedings*, 34, 331, Springer-Verlag, Heidelberg.

Prantzos, N. 2008. On the "Galactic Habitable Zone." *Space Science Reviews* 135, 313–322.

Race, M.S. 2015. Searching for extraterrestrial life: Are we ready? In *The Impact of Discovering Life Beyond Earth*, ed. S.J. Dick, Cambridge: Cambridge University Press, pp. 263–285.

Rampino, M.R. 2015. Disc dark matter in the galaxy and potential cycles of extraterrestrial impacts, mass extinctions and geological events. *Mon. Not R Astron. Soc.* 448, 1816–1820.

Randall, L. and Reece, M. 2014. Dark matter as a trigger for periodic comet impacts. *Phys. Rev. Lett.* 112, 16, 161301.

Scherer, K., Fichtner, N., Borrmann, T. et al. 2006. Interstellar-terrestrial relations: Variable cosmic environments, the dynamic heliosphere, and their imprints on terrestrial archives and climate. *Space Sci. Rev.* 127, 327–465.

Shaviv, N. 2002. The spiral structure of the Milky Way, cosmic rays, and ice age epochs on earth. *New Astronomy* 8, 39–77.

Shields, A.L., Ballard, S., Johnson, J.A. 2016. The habitability of planets orbiting M-dwarf stars. *Phys. Rep.* 663, 1–38.

Shostak, S. 2015. Current approaches to finding life beyond Earth, and what happens if we do. In *The Impact of Discovering Life Beyond Earth*, ed. S.J. Dick, Cambridge: Cambridge University Press, pp. 9–22.

Su, M., Slatyer, T.R., Finkbeiner, D.P. 2010. Giant gamma-ray bubbles from fermi-LAT: Active galactic nucleus activity or bipolar galactic wind? *Astrophys. J.* 724, 1044–1082.

Taleb, N.N. 2007. *The Black Swan: The Impact of the Highly Improbable*. Allen Lane.

Tarter, J. 2001. The search for extratterestrial intelligence (SETI). *Annu. Rev. Astron. Astrophys.* 39, 511–548.

Terrier, R., Ponti, G., Bélanger, G. et al. 2010. Fading hard X-ray emission from the galactic center molecular cloud Sgr B2. *Astrophys. J.* 719, 143–150.

Thomas, B.C., Engler, E.E., Kachelrieß, M., Melott, A.L., Overholt, A.C., Semikoz, D.V. 2016. Terrestrial effects of nearby supernovae in the early pleistocene. *Astrophys. J. Lett.* 826, 3–8.

Thomas, B.C., Jackman, C.H., Melott, A.L. et al. 2005. Terrestrial ozone depletion due to a Milky Way gamma-ray Burst. *Astrophys. J. Lett.* 622, 153–156.

Thomas, B.C., Melott, A.L., Field, B.D., Anthony-Twarog, B.J. 2008. Superluminous supernovae: No threat from η Carinae. *Astrobiol.* 8, 1, 9–16.

Thorsett, S.E. 1995. Terrestrial implications of cosmological gamma-ray burst models. *Astrophys. J. Lett.* 444, 53–55.

Vakoch, D.A. and Dowd, M.F. (eds.) 2015. *The Drake Equation: Estimating the Prevalence of Extraterrestrial Life through the Ages*. Cambridge: Cambridge University Press.

van Paradijs, J., Groot, P.J., Galama, T. et al. 1997. Transient optical emission from the error box of the γ-ray burst of February 28, 1997. *Nature* 6626, 686–689.

Vidal, C. 2015. A multidimensional impact model for the discovery of extraterrestrial life. In *The Impact of Discovering Life Beyond Earth*, ed. S.J. Dick, Cambridge: Cambridge University Press, pp. 55–75.

Wallner, A., Feige, J., Kinoshita, N. et al. 2016. Recent near-earth supernovae probed by global deposition of interstellar radioactive 60Fe. *Nature* 532, 69–72.

Ward, P. and Brownlee, D.E. 2000. *Rare Earth: Why Complex Life is Uncommon in the Universe*. Copernicus.

Wickramasinghe, J.T. and Napier, W.M. 2008. Impact cratering and the oort cloud. *Mon. Not R Astron. Soc.* 387, 153–157.

Wright, J.T. and Sigurdsson, S. 2016. Families of plausible solutions to the puzzle of Boyajian's Star. *Astrophys. J. Lett.* 829, 3.

Zubovas, K. and Nayakshin, S. 2012. Fermi bubbles in the Milky Way: The closest AGN feedback laboratory courtesy of Sgr A*? *Mon. Not R Astron. Soc.* 424, 1, 666–683.

CHAPTER 5

Space Debris

Risk and Mitigation

Camilla Colombo, Francesca Letizia,
Mirko Trisolini, and Hugh Lewis

CONTENTS

5.1 INTRODUCTION

The space surrounding our planet is densely populated by an increasing number of man-made space debris of which most derive from spacecraft abandoned in orbit after their operative life, upper stages and fragments generated by collision in space between space debris, or explosions (Flury 1995). The U.S. Space Surveillance Network (i.e., USSTRATCOM Two-Line Elements data set*) identified more than 23,000 artificial objects larger than 5 to 10 cm in size in *low Earth orbit* (LEO) and 30 cm to 1 m objects in *geostationary Earth orbit* (GEO), of which only 8% are operational spacecraft (see Figure 5.1). It is assumed that the population of objects larger than 1 cm is in the order of 750,000.† The forecast for the next decade is a growth of this debris population due to in-orbit explosions, material deterioration, and in-orbit collisions.

Space debris poses a threat to current and future spacecraft as fragments can collide at very high velocities and penetrate structures and damage satellite systems, leading to spacecraft anomalies and failures (McKnight 2016), such as, for example, the recent event of Sentinel-1A due to an impact from a micrometeoroid and orbital debris onto its solar array (Krag et al. 2017), or the catastrophic collision of the Cosmos 2251 satellite with the satellite Iridium 33 (Pardini and Anselmo 2017). Breakups generate a cloud of hundreds of thousands of fragments smaller than 10 cm, which are difficult to track or even to detect on a routine basis (the tracked object catalog covers objects larger than about 5–10 cm in LEO and 30 cm to 1 m at geostationary altitudes).

However, the first mitigation guidelines and practices for end-of-life disposal and debris reduction only date from a few decades ago, while it is now internationally recognized that, in the future, active control of the space debris environment will be necessary to allow safe space flight activities (U.S. Government 1997; Inter-Agency Space Debris Coordination Commitee 2002; Alby et al. 2004; United Nations Office for Outer Space Affairs 2010; International Standards Organisation 2011).

5.2 SPACE DEBRIS DISTRIBUTION AND EVOLUTION

Space debris are all man-made objects, including fragments and objects in Earth orbit or reentering the atmosphere, which are nonfunctional (Inter-Agency Space Debris Coordination Commitee 2002). Objects in the space

* USSTRATCOM Two-Line Elements data set: https://www.celestrak.com/NORAD/elements/ (last retrieved October 17, 2017).

† European Space Agency (ESA) website: http://www.esa.int/Our_Activities/Operations/Space_Debris/About_space_debris (last retrieved April 17, 2017).

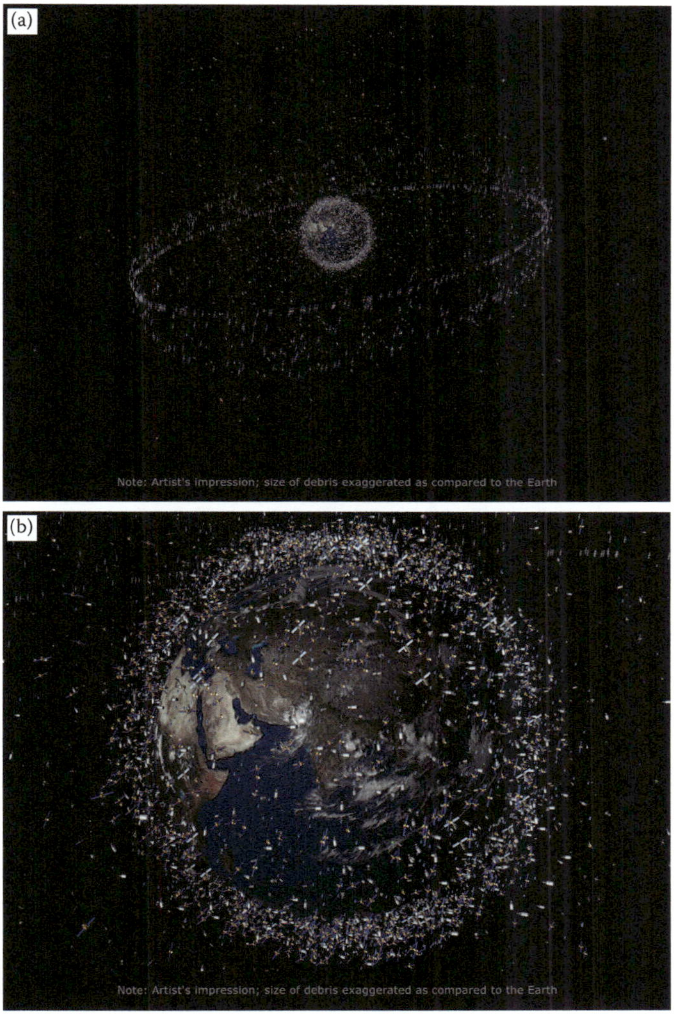

FIGURE 5.1 Objects around the Earth (artist's impressions). (a) Trackable objects in orbit around Earth, and (b) objects in low earth orbit (LEO)—view over the equator. European Space Agency. (©ESA.) (Note: The debris fields shown in the images are artist's impressions based on actual data. However, the debris objects are shown at an exaggerated size to make them visible at the scale drawn.)

environment include all the space objects from LEO to GEO, which are composed of payload (i.e., spacecraft), rocket bodies, mission-related objects (all objects dispensed, separated, or released as part of the planned mission), and fragmentation debris (fragments generated by space object breakup and anomalous event debris) (*NASA Orbital Debris Quarterly News* 2016).

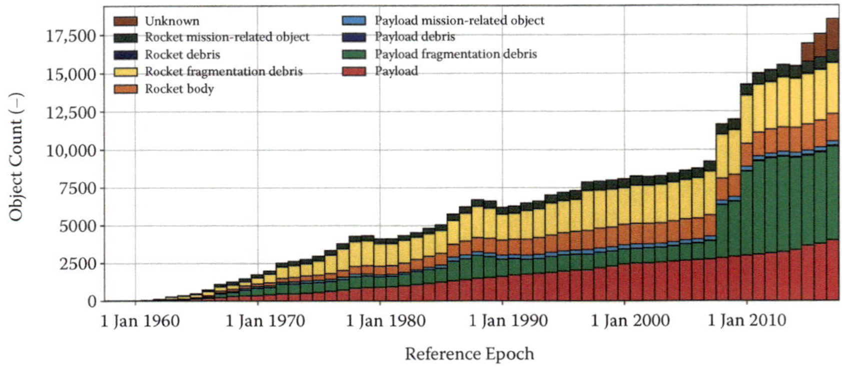

FIGURE 5.2 Evolution of the number of space objects in geocentric orbit by object class. (Produced by the European Space Agency Space Debris Office. From ESA Space Debris Office, 2017. ESA's Annual Space Environment Report. Produced with the DISCOS Database. https://www.sdo.esoc.esa.int/environment_report.)

Figure 5.2 shows the evolution of the space environment since the beginning of the Space Age (ESA Space Debris Office 2017). The evolution of cataloged objects in orbit is subdivided based on object classification (as it appeared in a space surveillance system during that year) as payload (i.e., objects launched in space to perform a specific mission, excluding rocket bodies), payload mission-related objects (objects released as product of a mission, such as cover of optical instruments), payload fragmentation debris (objects produced when a payload explodes or when it collides with another object), payload debris (space objects fragmented or unintentionally released from a payload as space debris but with no clear correlation with a particular event), rocket bodies (i.e., orbital stages of launch vehicles), rocket mission-related objects (i.e., shrouds and engines), rocket fragmentation debris (deriving from an explosion of a rocket body), rocket debris (space objects fragmented or unintentionally released from a rocket body as space debris but with no clear correlation with a particular event). In the case of the evolution of payloads and rocket bodies, the reported numbers are close to values known to exist in the environment from the mission design; in all other object classifications the amount of cataloged objects are almost certainly an underestimation and hence represent the lower limit for the true space environment (ESA Space Debris Office 2017).

Figure 5.3 reports the distribution of space objects in semimajor axis, eccentricity, and inclination for epoch January 1, 2013 for objects with

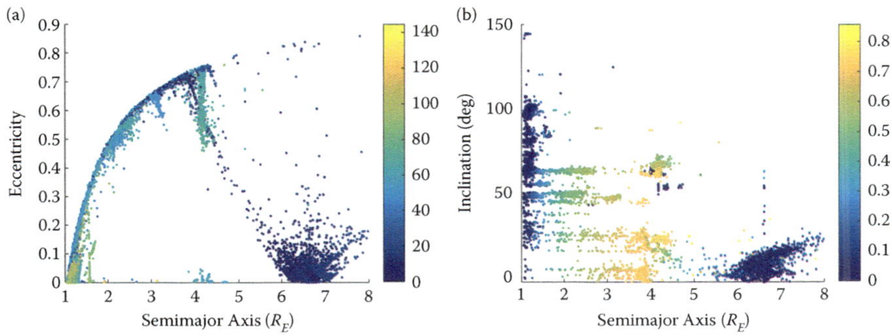

FIGURE 5.3 Distribution of total objects from LEO to GEO. (a) Semimajor axis-eccentricity, the color bar represents the inclination in degrees; (b) semimajor axis-inclination, the color bar represents the eccentricity. R_E is the radius of the Earth equal to 6371 km. (The data used for these figures were from MASTER, 2013. Meteoroid and Space Debris Terrestrial Environment Reference. https://sdup.esoc.esa.int. Retrieved October 1, 2017, by the ESA Space Debris Office.)

characteristic length larger or equal to 10 cm. The data were produced with MASTER 2013 by the ESA Space Debris Office (MASTER 2013). As can be seen, many objects are concentrated in particular slots of semimajor axis and inclination. This is particularly visible in Figure 5.4, which represents the spatial density of objects between 0 and 3000 km altitude calculated as the number of objects in each orbit bin, considering a dimension of orbit bin of 20 km. The population of space objects is not uniformly distributed but presents a peak in the LEO region between 800 and 900 km, corresponding to the altitude selected by many remote sensing missions who chose a Sun-synchronous orbit because of the stable sun-lighting conditions which favors multiple observations. The main contribution to this peak is due to fragmentation debris and sodium-potassium droplets released from orbital nuclear reactors operated in space before the end of the 1980s (Wiedemann et al. 2017). A second peak is present between 1400 and 1500 km altitude.

The orbit evolution of space debris is governed by the Earth's gravity potential but also by the effect of orbit perturbations due to atmospheric drag, solar radiation pressure, and third body luni-solar perturbations. As the ratio between the cross area and the mass of each object is different (see Figure 5.5), they have a different orbit evolution under the effects of atmospheric drag and solar radiation pressure which overlap the effects of luni-solar perturbation and nonuniform distribution of the Earth's gravity field.

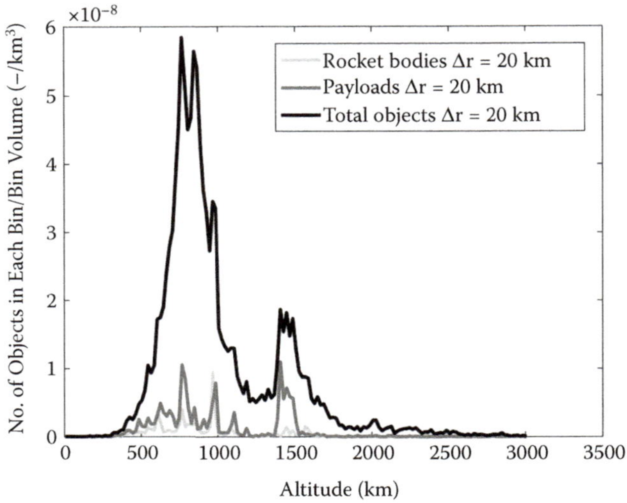

FIGURE 5.4 Spatial density as function of the semimajor axis. (The data used for this figure were produced by MASTER 2013 by the ESA Space Debris Office. From MASTER, 2013. Meteoroid and Space Debris Terrestrial Environment Reference. https://sdup.esoc.esa.int. Retrieved October 1, 2017.)

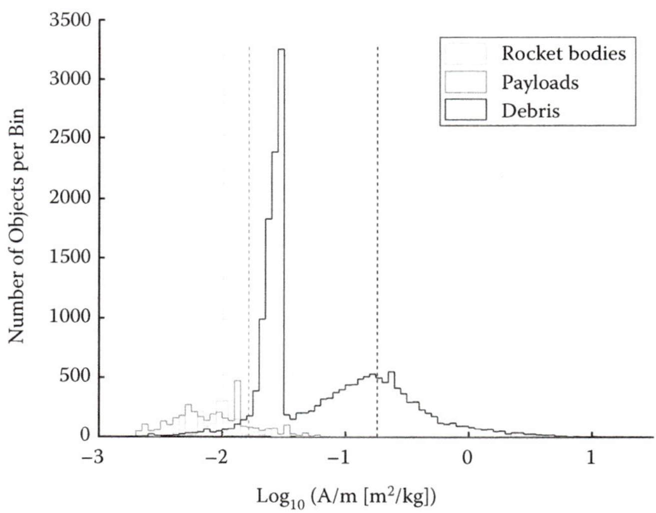

FIGURE 5.5 Number of objects in each bin versus their value of area-to-mass ratio in m²/kg. Dashed lines: mean value. (The data used for this figure are from MASTER, 2013. Meteoroid and Space Debris Terrestrial Environment Reference. https://sdup.esoc.esa.int. Retrieved October 1, 2017.)

Space debris models provide the distribution of objects in space, their physical characteristics (size, optical properties, mass, rotational motion), and their movement. The approach to investigate the evolution of orbital debris and spacecraft population relies on codes which numerically simulate the long-term evolution of the space object population over 100 to 200 years under the effects of orbit perturbations (Rossi et al. 1997). Sources due to new launches, fragmentation events (explosions and collisions), new debris release, spacecraft degradation and sinks due to active debris removal (i.e., intentionally remove via a dedicated mission a selected debris object from the environment), and end-of-life mitigation (Sdunnus et al. 2004) are taken into account. The space object population is propagated statistically (i.e., characterization through a set of representative objects of the whole population), or deterministically (i.e., following the dynamics of each member). These models are applied to assess the risk and damage, in order to predict the frequency of collision avoidance maneuvers for operational spacecraft and to evaluate the effectiveness of current debris mitigation measures. In particular, they are used to study the long-term effect of different strategies of spacecraft disposal at the end of the operational life or to evaluate the effectiveness of active space debris removal concepts (Anselmo et al. 2001).

The future projection of the space debris population can be studied via numerical simulations based on long-term evolution models, which implement profiles of the future launch traffic, solar activity, and different levels of compliance to space debris mitigation strategies. There are several uncertainties associated to these models (Dolado-Perez et al. 2015): The solar and geomagnetic activity, which greatly influence the value of the atmosphere density at a given altitude and time, and therefore the object decay rate is difficult to predict (Vallado and McClain 2007). On top of this, the breakup model used to determine the fragments' velocity, mass, and size distribution following a collision or an explosion is based on a limited number of ground tests and on-orbit breakups characterized by orbital tracking (Johnson et al. 2001; Krisko 2011). Additionally, collision prediction algorithm and Post Mission Disposal compliance rates cannot be predicted, thus all of these factors result in uncertainties in the future space object population growth and in the prediction of the number of objects in the LEO to GEO region. The results of these models can be used to analyze the effectiveness and robustness of mitigation measures in constraining the growth of the space debris population (Inter-Agency Space Debris

Coordination Commitee 2002) rather than to predict the actual future evolution (Dolado-Perez et al. 2015).

5.3 RISK RELATED TO SPACE DEBRIS

The risk related to space debris can be divided into two categories; the first is the on-orbit risk due to the possibility of collisions and explosions. The time evolution of the number of objects in Earth orbit, officially cataloged by the U.S. Space Surveillance Network (*NASA Orbital Debris Quarterly News* 2016), clearly shows the effects of fragmentations such as those generated from the Chinese FengYun 1 C antisatellite test in 2007 (Pardini and Anselmo 2009) and the collision between Iridium 33 and Cosmos 2251 in 2009 (Pardini and Anselmo 2017), which resulted in a sudden and major increase in the number of space debris. While primary causes for spacecraft anomalies are due to design issues, another cause has to be the collisions of spacecraft with space debris, especially at altitudes with a high number of cataloged objects (McKnight 2016). A quantitative estimation of this on-orbit risk can also be appreciated by looking at the number of collision avoidance maneuvers measured at a given orbit altitude, which depends on the probability of collision. Just as an example, the probability of collision over one year (in 2013) for a satellite with a 20 m^2 surface area at an orbital altitude close to the one of the SPOT (Satellite Pour l' Observation de la Terre) satellite (i.e., 825 km) was of 2×10^{-4} for objects larger than 10 cm, 3×10^{-3} for objects larger than 1 cm, 0.5 for objects larger than 1 mm, and 1 for objects larger than 0.1 mm (Durrieu and Nelson 2013). Moreover, in case of a warning for a possible collision event, assessment at the ground control level needs to be made in a timely manner to determine whether to actually perform a collision avoidance maneuver, as this could introduce a secondary high-risk event, a change on the ground track, and a momentary suspension of the mission service during execution of the maneuver, which need to be determined taking also spacecraft constraints into account (Symonds et al. 2014). Space agencies and spacecraft operators receive conjunction notifications from the Joint Space Operations Center (JSpOC) while the European Space Agency (ESA)'s Space Debris office provides an operational service for the assessment of collision risks of ESA satellites (Flohrer et al. 2009). Research is also devoted to the development of onboard detectors for the in situ measurement system of submillimeter class microdebris. For example, the Japan Aerospace Exploration Agency (JAXA) has tested on board the International Space Station an in situ Space Debris Monitor (SDM) which uses conductive (resistive) strip lines for microdebris detection.* If we consider that the number of launches per year is increasing (e.g., between 2010 and 2016 the average number of launches of spacecraft with a mass less than 1000 kg was 83 satellites per year [Colombo et al. 2017]), due to the current trend in spacecraft miniaturization, the on-orbit risk will increase in the future.

The second factor associated to space debris risk is the casualty risk on ground, every time an uncontrolled or semicontrolled object reenters the Earth's atmosphere the aero-thermodynamics interaction with the atmosphere causes the demise of a fraction of the overall mass of the spacecraft. However the surviving components, which are reduced in fragments of different size and mass, can pose a risk for people and properties on the ground. Related to the reentry of space objects, recent studies are also attempting to estimate their impact on the pollution of Earth's environment through environmental impact assessments (Durrieu and Nelson 2013; BIO Intelligence Service 2014).

5.3.1 On-Orbit Risk

As mentioned in the previous section, space debris has an impact on the operation of satellites. Thus, its presence becomes a hard constraint for the exploitation of space and different measures have been adopted to limit its growth in the long term. These include the passivation of rocket bodies (to limit the risk of in-orbit explosions) and the definition of protected orbital regions, which should be left clear at the end of a mission. Some of these measures result from the observation that the long-term evolution of the space debris environment is highly affected by the fragmentation of large intact objects (Rossi et al. 2015). A fragmentation can be caused by explosion (e.g., due to an onboard failure) or by a collision with another object. In both cases, a cloud of fragments is generated: The cloud, initially dense and localized, spreads under the effect of different forces, and thus a fragmentation can affect objects in different orbital regimes.

For this reason, different metrics have been proposed to rank the risk posed by space objects depending on the consequences of their potential fragmentation on the space environment. The purpose of these analyses is twofold. First, to obtain a better insight on the critical parameters which have the largest influence on the space debris evolution. This is

* JAXA Space Debris Monitor: http://www.kenkai.jaxa.jp/eng/pickup/sdm.html (last retrieved October 1, 2017).

useful to support both the licensing of new missions and the analysis of the sustainability of the current level of compliance to the space debris mitigation guidelines. Second, the output of these rankings could lead to the identification of potential candidates for active debris removal missions: in such a scenario, it would be important to decide which spacecraft should be removed first to have the largest global beneficial effect.

Several formulations have been proposed to quantify the risk of an orbit fragmentation, focusing in different ways on its two components, probability and severity. Liou and Johnson in 2009 were among the first to propose an indicator to summarize in a single value the relationship between a space object and the debris environment. In particular, their indicator multiplies the collision probability due to the background debris population and the object mass, which can be seen as an indicator of the severity of the resulting fragmentation. The main application for this indicator was for the identification of candidates for active debris removal missions.

Yasaka (2011) can be seen as the first to propose a debris index specifically designed to guide the licensing process and to drive the design of spacecraft and upper stages, as well as to facilitate the understanding of the debris issue by the general public. For this reason, the formulation by Yasaka (2011)—as opposed to that from Liou and Johnson in 2009—is *analytical*, meaning that the value of the index can be computed directly from the knowledge of some specific quantities related to the analyzed object, but *ad hoc* simulations of the evolution of the debris environment are not required. The proposed formulation is as follows

$$I_{\text{deb}} = \alpha M A \Phi(h) T_{\text{orb}} \tag{5.1}$$

where α indicates the number of fragments created per unit mass by a *catastrophic* collision,* M is the mass of the object, A its cross-sectional area, $\Phi(h)$ the flux of debris at the altitude h, T_{orb} the orbital lifetime of the object. The first two terms refer to the expected severity of the potential fragmentation, whereas the other terms refer to the likelihood of the event. A similar formulation was later proposed also by Utzmann et al. (2012).

While analytical indices are very effective in highlighting the main drivers of the space debris evolution, they are always based on some simplifications and cannot capture phenomena such as feedback effects, that

* A catastrophic collision is an event where the two colliding objects are completely destroyed.

is, the possibility that the generated fragment cloud can trigger additional collisions within the debris population. For this reason, other authors have based their indices on extensive simulations of the debris environment. For example Bastida et al. in 2013 used the output of the simulations of the debris environment in terms of the generated catastrophic collisions to develop a criterion for the selection of candidates of active debris removal missions. Another index was proposed by Kebschull et al. in 2014 to evaluate the environment criticality, based on the computation of the change in the collision probability for the whole debris population due to the fragmentation of a selected object. A catastrophic collision is simulated at different time instants within the considered time window and its effect on the global debris population is estimated by applying an analytical model of the debris evolution. Also, in this case the proposed index of criticality takes into account both the consequences of the fragmentation and its probability of happening.

Based on a the results of a numerical sensitivity analysis of the effect of collisions in several selected cases in LEO and GEO with different values of colliding mass, orbital parameters, and object type, Rossi et al. in 2016 (Rossi et al. 2015) defined a criticality index for LEO based on four key elements: environmental dependence, lifetime dependence, mass, and inclination. All these factors are combined in one index called the Criticality of Spacecraft Index:

$$I_{\mathrm{deb},\Xi} = \frac{M}{M_0}\frac{D(h)}{D_0}\frac{T_{\mathrm{orb}}(h)}{T_{\mathrm{orb}}(h_{1000})}\frac{1+k\Gamma(i)}{1+k} \tag{5.2}$$

where the mass of the analyzed spacecraft is M, the spatial density of objects at the orbital altitude h is $D(h)$, the expected orbital lifetime of the object given its orbital altitude h is $T_{\mathrm{orb}}(h)$ are normalized with respect to some reference values. The last multiplying factor in Equation 5.2 with $k = 0.6$ and $\Gamma(i) = (1 - \cos(i))/2$, considers the effect of the inclination. A similar formulation was proposed by Anselmo and Pardini in 2015 and Anselmo and Pardini in 2016 with the difference that the space debris flux $\Phi(h,i,M)$ is used instead of the debris density. $\Phi(h,i,M)$ is function of altitude, inclination, and mass as only the flux of objects able to trigger a catastrophic collision is considered.

$$I_{\mathrm{deb,RN}} = \left(\frac{M}{M_0}\right)^{1.75}\frac{\Phi(h,i,M)}{\Phi_0(h,i,M)}\frac{T_{\mathrm{orb}}(h)}{T_{\mathrm{orb}}(h_0)} \tag{5.3}$$

Also, in these two cases (Equations 5.2 and 5.3) one can recognize the combination of terms related to the probability and the severity of potential fragmentation.

A different approach was followed by Lewis et al. in 2013 who proposed an Environmental Impact Rating (EIR) system to evaluate how spacecraft design and operation can impact the long-term debris environment. The EIR system is composed of three main components:

- A debris score related to the density of debris along the orbit of the spacecraft.
- A capacity score, which assesses how the selected design or operational baseline can reduce the future production of debris.
- A health score representing how the spacecraft affects the health of the orbital region where it operates.

The EIR system was implemented as a web tool,* which generates a graphical output similar to the very familiar European energy label. This representation is particularly interesting because it suggests that, as in the case of household appliances, design/operational improvements bring benefits both for the single operators and to the whole environment.

Following this idea Letizia et al. in 2016 proposed an index to measure the Environmental Consequences of Orbital Breakups (ECOB); this index, differently from the others, ranks space objects considering the effect of their fragmentation on other operative satellites. The choice of looking at the effects on operational satellites is because this can be more easily connected to the cost to satellite operators due to fragmentations (*private cost*). In addition, the collision risk for operational satellites may also be seen as an indicator of the availability of future access to space (*shared cost*) because the orbital regions with most operational satellites are the ones that offer a privileged point of view for Earth observation. For example, this is the case of sun-synchronous orbits, which allow the Earth to be observed with constant illumination conditions. Therefore, they are also expected to be an important asset in the future.

---

* The EIR web tool is available online at http://www.fp7-accord.eu/rating (last retrieved August 26, 2016).

The National Aeronautics and Space Administration (NASA) breakup model (Johnson et al. 2001; Krisko 2011) is used to simulate the breakup and the generation of the fragment cloud. An analytical propagation method, CiELO (debris Cloud Evolution in Low Earth Orbit) was developed to describe how the fragmentation cloud evolves in time and the resulting collision probability for objects crossing the cloud (Letizia et al. 2015). This approach is used to quantify the severity of fragmentations, whereas historical data on explosions and ESA flux model MASTER are used to estimate the probability of fragmentation. The next sections will explain more in detail how the index is computed.

5.3.1.1 Evaluating the On-Orbit Risk

As highlighted in the previous section, many methods have been proposed to quantify the risk caused by space debris in orbit. Two main aspects can identify the interaction of a spacecraft, during its operational or end-of-life phase, with the space debris environment. The *probability* of fragmentation caused by the space debris environment on the analyzed mission (i.e., probability of collision) and from stored energy onboard (i.e., probability of explosion). The probability of collision is a function of the flux of space debris, the operational orbit of the object and its trajectory evolution, the capabilities of collision avoidance maneuvering by the object under analysis (therefore its object type: spacecraft, rocket body, etc.), and its cross-sectional area. The *severity*, instead, measures the consequent effect on the space environment of the analyzed mission scenario. Therefore, it is function of the mass of the object, of the characteristics of the breakup (i.e., collision or explosion), and of the orbit where the breakup occurs.

In summary, all these elements are collected in the proposed index based on the risk associated to fragmentation events (Letizia et al. 2017), whose structure is.

$$I_{debris\ risk} = p_e \cdot e_e + p_c \cdot e_c \tag{5.4}$$

where p_c is the probability of a collision happening, and e_c measures the effects of the collision on operational satellites, p_e is the probability of an explosion happening, while e_e measures the effects of the explosion on operational satellites. A thorough description of the index is given in Letizia et al. 2017; a summary is given in the next paragraphs.

5.3.1.1.1 On-Orbit Collisions The probability of collision p_c is computed through the kinetic gas theory, so that the cumulative collision probability is written as

$$p_c = 1 - \exp(-\rho(h)\Delta v A \Delta t) \tag{5.5}$$

where ρ is the debris density at the spacecraft orbit, Δv is the collision velocity, A the collision area, and Δt is a fixed time interval. For the debris index, an appropriate value of Δt should be chosen. The collision velocity of a given spacecraft orbiting through the space debris environment is calculated from ESA MASTER simulations, building a grid of the most likely impact velocity for a spacecraft at a given semimajor axis and inclination on a circular orbit. Similarly, also the term ρ is obtained from MASTER, considering not only the dependence on the altitude but also on the fragment mass. This is done because, as in Anselmo and Pardini 2015, we are interested only in the number of objects in the debris population able to cause the complete destruction of the target object.

The effect of the collision e_c is measured by the resulting increase in the collision probability for operational satellites (Letizia et al. 2016). A set of targets representative of the whole population of operational satellites is defined based on the distribution of the cross-sectional area. A grid in semimajor axis and inclination is introduced, and a representative target for each cell with the highest cumulative cross-sectional area is selected. Figure 5.6 shows an example of the grid used for the simulation while the representative targets are indicated with a circle. Once the target set is defined, the effect of fragmentation can be evaluated. A key point of the suggested approach is not to compute the index only for specific objects, but rather to study its dependence on parameters such as orbit altitude, inclination, and spacecraft mass.

A fragmentation is triggered for each cell in the same grid in semimajor axis and inclination, and for each event the resulting cloud of fragments is propagated through a density-based approach. The collision probability on each of the representative targets is computed with the same expression as Equation 5.5, where now ρ is the spatial density of the fragmentation cloud at the spacecraft altitude, Δv is the relative velocity between the target and the fragments in the cloud, A is the cross-sectional area, and Δt is the time span used for the computation.

The effect on each target is summed and modified through a weighting factor to take into account that each representative target is associated with a different share of the total spacecraft area distribution. The resulting

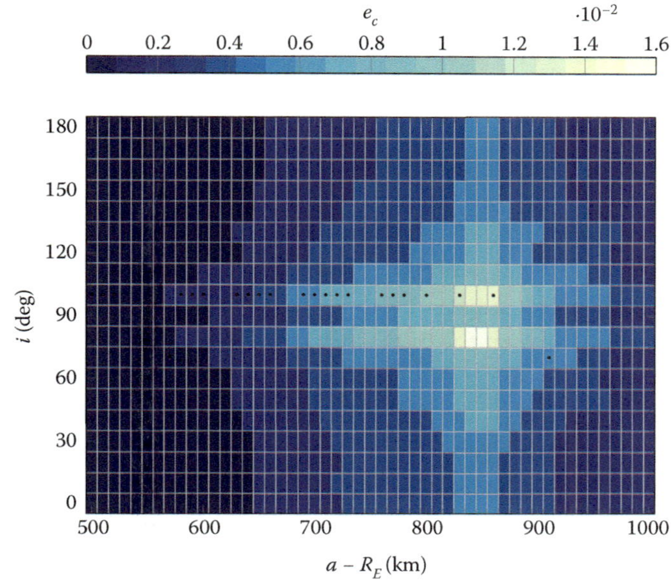

FIGURE 5.6 Variation of e_c with the semimajor axis and the inclination of the parent orbit of the fragmentation. The gray dots refer to the reference targets used for the simulation.

collision probability, shown in Figure 5.6, refers to a set value of the mass of the fragmenting object, but then this value can then be rescaled with a power law to consider different masses of the fragmenting object (Utzmann et al. 2012; Letizia et al. 2016). One of the advantages of studying the index dependence on the orbit parameters (rather than only evaluating single spacecraft) is that maps such as Figure 5.6 show clearly which are the most critical orbital regions. In addition, the possibility of analytically rescaling the results to account for the mass means that no additional simulations are required and that only the map in Figure 5.6 is needed to compute the term e_c for any spacecraft.

5.3.1.1.2 On-Orbit Explosions An analytical expression for the probability of explosion p_e can be derived by analyzing statistical data, as a function of the time elapsed between the launch of the object and its fragmentation. Two different curves can be derived in this way, distinguishing between payloads and rocket bodies (Letizia et al. 2017).

For the term e_e, the same approach as for e_c is followed. With respect to collisions, explosions are expected to produce larger fragments with lower speed (Johnson et al. 2001). The NASA breakup model is used also in this case to generate the corresponding debris cloud. As in the previous case, an explosion is triggered in each cell of a grid in semimajor axis and inclination, thus the resulting fragment cloud is propagated and its effect measured on the defined representative targets.

5.3.1.1.3 Index Computation By putting all the terms together, a map such as the one in Figure 5.7 is obtained, where the light-colored areas are the ones with the highest fragmentation risk and the white markers indicate the top 25 payloads with the highest value of the index.

This index can also be evaluated on a mission profile to distinguish between spacecraft that do and do not implement end-of-life disposal. When a disposal strategy is implemented, the spacecraft leaves its slot at the end of the mission and it is either moved toward higher altitudes or toward the Earth to reenter the atmosphere and burn up. In both cases, the fragmentation risk associated to the object will be largely reduced. To consider this aspect, the calculation of the debris risk $I_{debris\ risk}$ is carried out integrating the spacecraft trajectory and computing the value of the

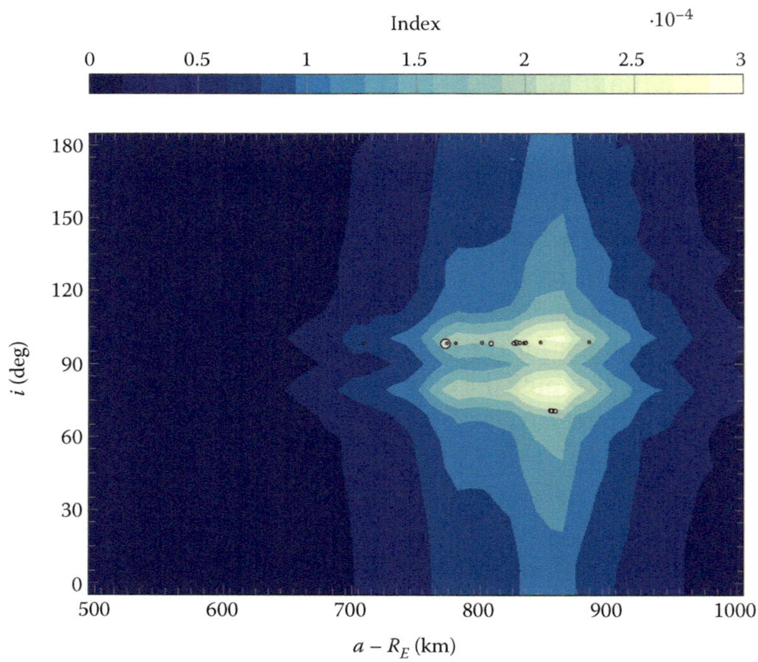

FIGURE 5.7 Variation of the index with the orbital parameters.

terms p_c, e_c, p_e, e_e at each time step. In this way, different missions (or different disposal options) can be compared by looking at the cumulative debris index over the mission profile.

5.3.2 On-Ground Risk

5.3.2.1 Casualty Risk Definition

When a spacecraft reenters the atmosphere, the severe heating environment destroys a significant fraction of the overall mass of the spacecraft, however, about 10% to 40% of the initial mass is expected to survive the reentry and impact the Earth's surface (Patera 2008). The surviving components can possess sufficient mass and energy to pose a risk for people on the ground. This can be evaluated using the concept of casualty risk expectation, which is the probability of the surviving objects to impact people on the ground. Such probability is a function of the characteristics of the surviving objects such as the cross-sectional area, the impact energy, as well as the impact location on the ground, the population density at the impact point, and epoch. In its most general form, the casualty risk expectation (E_c) can be expressed as (Anselmo and Pardini 2005):

$$E_c = P_i N \frac{A_c}{A_f} \tag{5.6}$$

where N is the total population in a region of area A_f, P_i is the probability of impact in the region, and A_c is the total effective casualty area of the components reaching the ground. The total effective casualty area is defined as the summation on all the n individual fragments reaching the ground, which takes into account both the contribution of the cross section of the surviving debris A_i and the projected cross-sectional area of a standing human A_h:

$$A_c = \sum_{i=1}^{n} \left(\sqrt{A_h} + \sqrt{A_i} \right)^2 \tag{5.7}$$

The prediction of the impact location of the surviving fragments is highly uncertain for long-term reentry predictions because of the uncertainty in atmospheric drag and reentry time (Klinkrad et al. 2006). Only for short-term predictions in the order of few days is it possible to better predict the location of the reentry footprint. The computation of the casualty risk can thus be treated differently in case of long-term and short-term predictions. When long-term predictions are considered, the related uncertainties allow for the consideration of a uniform distribution of the reentry position within the orbit plane. Moreover, at low Earth orbit, the equatorial bulge causes the orbit plane to precess about the Earth's axis, thus changing the right ascension of the ascending node. Therefore, given the uncertainty in atmospheric drag, the right ascension of the ascending node of the final orbit can also be considered randomized. In addition, the uncertainty in the reentry time randomizes the longitude range of the reentry footprint. With these considerations for long-term predictions, the probability of ground impact can be expressed with an impact probability density function, which depends only on the orbit inclination and on the latitude of the point of impact (Patera 2008). This impact probability density function is coupled with the population density to derive the casualty expectation integral, which is used to calculate the casualty expectation as a function of the orbit inclination:

$$E_c = \frac{A_c}{\pi}\int_0^{\pi/2} \rho_p\left(-\sin(i)\sin(\theta)\right)d\theta + \frac{A_c}{\pi}\int_0^{\pi/2} \rho_p\left(\sin(i)\sin(\theta)\right)d\theta \tag{5.8}$$

where ρ_p is the population density, i is the orbital inclination, θ is the angle between the ascending node and the reentry position in the orbit plane, and A_c (see Equation 5.7) is the casualty area of the fragments. The integrals are separated to take into account the upper and lower Earth hemispheres. The term inside the integral is simply the population density as a function of the latitude λ, as $\sin(\lambda) = \sin(i)\sin(\theta)$.

On the other end, when the predicted orbital lifetime of an uncontrolled reentry drops below a few days, it is possible to have a more accurate prediction of the reentry and thus a more detailed risk assessment can be performed (Klinkrad et al. 2006). In this case, the ground impact corridor needs to be analyzed using a 2-D impact probability density function.

$$E_c = A_c \sum_{n=1}^{N}\sum_{m=1}^{M}\left(p_{i,2\sigma}\right)_{n,m} \times (\acute{\rho}_p)_{n,m} \tag{5.9}$$

where N and M represent the along track and cross track discretization of the ground swath area during the reentry, $(\acute{\rho}_p)_{n,m}$ is the average population density in the sampled ground swath area, and $(p_{i,2\sigma})_{n,m}$ is the local impact probability, which is defined as

$$(p_{i,2\sigma})_{n,m} = PDF_{2\sigma}(s_{xn}, s_{ym}) \times \Delta s_x \times \Delta s_y \tag{5.10}$$

where *PDF* is a 2-D probability density function, which takes into account the dispersion of the fragments on the ground and can be usually approximated with a Gaussian distribution. sx_n and sy_n are the predicted impact point around which the Gaussian is centered, and Δs_x and Δs_y are the along track and cross track length of the sampling of the ground swath area. The characteristics of such PDF may be obtained from a reentry analysis using destructive reentry software.

5.3.2.2 Software for Spacecraft Reentry Analysis

To assess the compliance of a spacecraft with the casualty risk threshold set as a requirement (Inter-Agency Space Debris Coordination Commitee 2002; ESA 2008) to 10^{-4}, reentry simulation software is used to analyze the atmospheric disposal of forthcoming missions. A variety of software is available to perform such an analysis. Destructive reentry codes can be divided into two categories: spacecraft oriented codes and object-oriented codes (Lips and Fritsche 2005).

Spacecraft oriented codes model the complete spacecraft using a triangular mesh description of the spacecraft structure. Aerodynamics and aerothermodynamics are analyzed for the effective spacecraft geometry. Breakup events and fragmentations are actually calculated assessing the thermal and mechanical loads on the spacecraft structure. An example of spacecraft oriented software is SCARAB (Spacecraft Atmosphere Reentry and Aerothermal Breakup; Koppenwallner et al. 2004). In SCARAB, the geometry of the spacecraft is modeled as a panelized structure, with each panel having geometrical and mechanical properties attached to it. The material database takes into account temperature-dependent properties. The reentry trajectory is analyzed with a complete six-degree-of-freedom dynamics, thus considering also the attitude motion of the spacecraft. Fragmentation events and separation of the spacecraft into multiple parts are also considered. The aerodynamics and aerothermodynamics analyses are based on local inclination methods where the pressure distribution and the heat rates are functions of the local surface inclination with respect to the external flux. The thermal analysis is based on a two-dimensional heat conduction model; however, heat radiation between each part of the spacecraft is also considered.

In object-oriented codes, on the other hand, the spacecraft structure and components are schematized using elementary shapes such as spheres, cylinders, boxes, and flat plates and the trajectory is propagated using a three-degree-of-freedom dynamics. The shapes used are simple but they

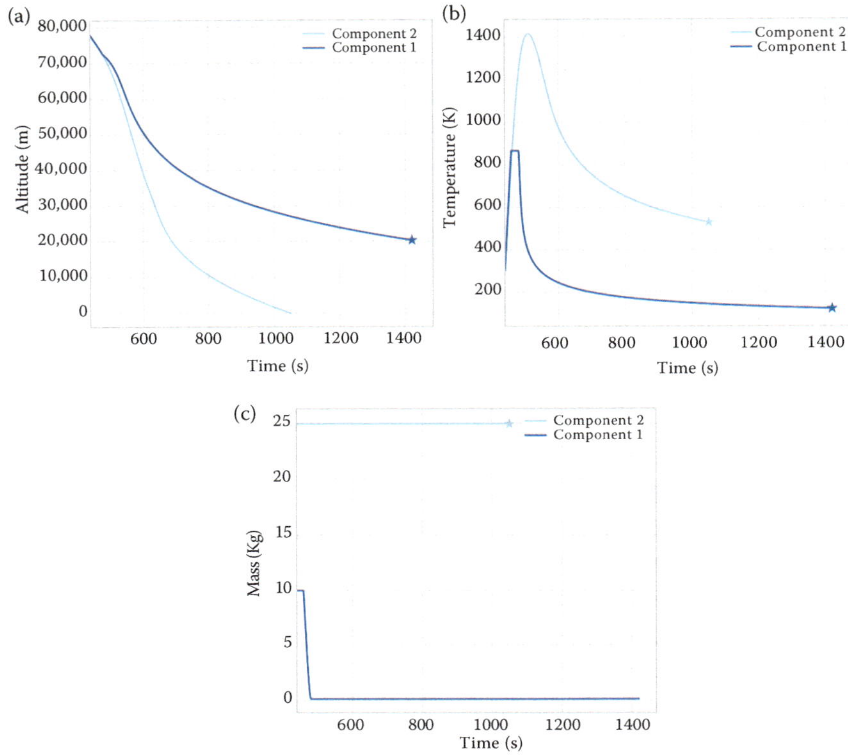

FIGURE 5.8 (a) Altitude, (b) temperature, and (c) mass profile evolution over time for two different components after the breakup event occurred at an altitude of 78 km.

can still adequately represent common spacecraft components like tanks and battery assemblies by assigning them proper dimensions and materials of components. The spacecraft architecture is built by first creating a parent object which represents the main spacecraft structure; then all the internal components are modeled using these elementary shapes. This kind of software usually assumes a fixed breakup altitude in the range 75 to 85 km and a predefined attitude motion of the components (usually random tumbling). Object oriented codes rely on attitude averaged drag coefficients and heat rates, together with a correction factor, which takes into account the shape of the object considered, in order to compute the drag and the heat flux on the reentering objects. For example Figure 5.8a shows the altitude profile evolution over time for two different components after a breakup event occurred at 78 km of altitude (Trisolini et al. 2015,

2016). Component 1 is an aluminum box with a side length of 0.5 m, while the second component is a stainless steel sphere of radius 0.5 m. It is possible to observe that the box has a quite shallow trajectory and after about 1400 s has not yet reached the ground. This is due to the mass loss during the reentry, making the box lighter and causing it to travel further. The simulation is actually stopped at around 1400 s as the component has reached an energy below the 15 J threshold and is no longer considered dangerous for the people on the ground. The sphere, instead, reaches the ground after just about 1050 s, as it is heavier and does not suffer from any demise. As can be seen from the temperature profile as function of time in Figure 5.8b, the aluminum box reaches the melting temperature and loses most of its mass (see Figure 5.8c), whereas the steel sphere does not reach the melting temperature and stays intact.

5.3.2.3 Casualty Risk Computation

For the calculation of the casualty risk during the first stages of mission design, object-oriented codes are usually used. If an object-oriented analysis suggests that the spacecraft may not be compliant, then a more detailed analysis is performed to better assess the compliancy (with a spacecraft oriented code or a more refined object-oriented code). On the ESA side, the Debris Risk Assessment and Mitigation Analysis (DRAMA) software is used to assess the compliance of space missions with a ground casualty risk limiting threshold of 10^{-4} (Inter-Agency Space Debris Coordination Committee 2002; ESA 2008). The DRAMA module includes a module that performs the reentry analysis, whose inputs are used to compute the casualty risk (Martin et al. 2005; Gelhaus et al. 2013). The reentry trajectory conditions from the orbital simulation down to 120 km can be used as initial conditions to propagate the trajectory with an object-oriented code down to 78 km (assumed as the breakup altitude), and applying biases to the atmospheric density. For the propagation below 78 km, no further density biases are applied so that one set of initial conditions at 78 km is produced for each atmospheric bias. To assess the risk to the population, a rectangular ground impact corridor is assumed with a fixed 2σ cross-track extension of ± 40 km. The along-track extension is defined by the trailing and leading impact point of each surviving fragment footprint. The trailing edge corresponds to the $+20\%$ density bias, whereas the leading edge to the -20% density bias, or the first trajectory that reaches the ground without demising. For every surviving object, the casualty area and the geodetic impact coordinates are provided as a function of the applied density biases.

The ground risk computation can be computed using the biased reentry simulation and the population density, which is defined on a latitude λ and longitude ϕ grid with a resolution of 15'. An exponential growth of the population in time t (expressed in years) is assumed since 1994.

$$\rho_p(\lambda,\phi,t) = \rho_p(\lambda,\phi,1994)\exp\left(\frac{t-1994.5}{59.63}\right)$$

The ground risk computation depends on the reentry forecast of the mission (i.e., short-term or long-term prediction). For long-term predictions, as the reentry location on the orbit is unknown, a uniform impact probability is assumed for a given orbit inclination $(P_i)_k = (\Delta s_x)_k/(2\pi R_E)$, where R_E is the Earth's radius, Δs_x is the along-track extension of the rectangular ground impact corridor, and k is the number of bins in which the reentry corridor is subdivided. For the same reason, the population density is averaged in longitude $\bar{\rho}_p(\lambda, \mathrm{t})$. In addition, due to the symmetry of the problem, a single orbit is used as the analysis interval. The expression for the corresponding casualty risk is then

$$E_c = 1 - \prod_{j=1}^{J}(1 - E_{c,j})$$

where the total casualty risk calculated for j = 1, .., J surviving objects is constructed from each individual contribution $E_{c,j}$

$$E_{c,j} = \sum_{k=1}^{N}(P_i)_k(\bar{\rho}_p)_k\hat{A}_c$$

where $\hat{A}_c$ is a mean casualty area, which is obtained from a weighted average over all possible along-track impact locations, with weights provided by the impact probability density function $(\mathrm{PDF}_{2\sigma})_k$ as a function of the impact location:

$$\hat{A}_c = \sum_{k=1}^{N}(PDF_{2\sigma})_k(\Delta s_X)_k(A_c)_k$$

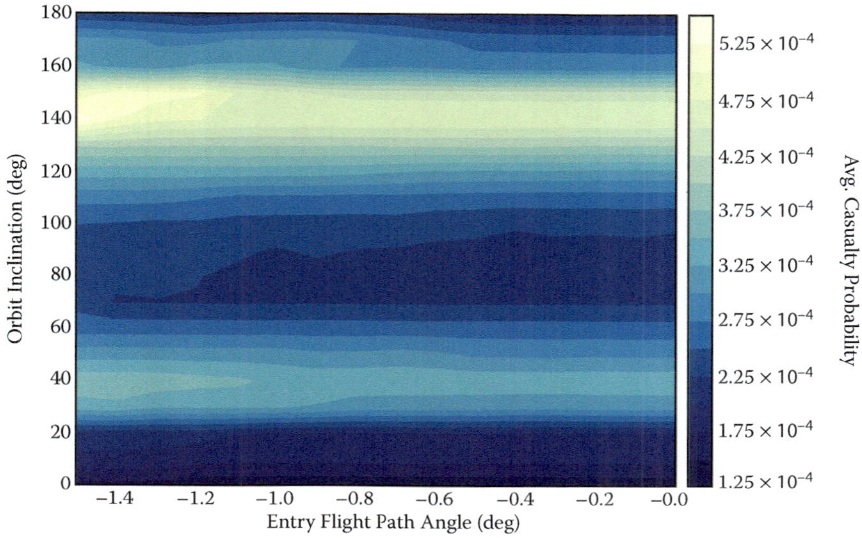

FIGURE 5.9 Casualty risk as function of orbit inclination and entry flight path angle for a 7.3 km/s entry velocity.

As an example, Figures 5.9 and 5.10 show a sensitivity analysis on different reentry conditions onto the casualty area and the impact masses, which reach the ground, performed with several DRAMA simulations. Figure 5.9 shows a map of the casualty risk as a function of the entry flight path angle and of the orbit inclination for a fixed relative velocity of 7.3 km/s. The impact mass increases moving from direct to retrograde orbits, and becomes lower for flight path angles around $-0.5°$. The casualty risk follows more closely the population distribution on the Earth, where the highest concentrations can be found at intermediate latitudes ($\pm 45°$). The inclination thus influences the casualty risk most, whereas the flight path angle produces less significant effects, as the casualty risk analysis performed uses a longitude averaged population density. Figure 5.10 shows the variation of the casualty risk as a function of the entry velocity and flight path angle for specific values of the orbit inclination. It is evident that for moderate velocities the higher the entry velocity the higher the demise, as the heat load on the spacecraft will be greater. However, the higher the relative velocity the greater the chance the spacecraft will not reenter (dark gray areas), especially for direct orbits. The flight path angle influence is instead related to the orbit inclination. For the 30° inclination

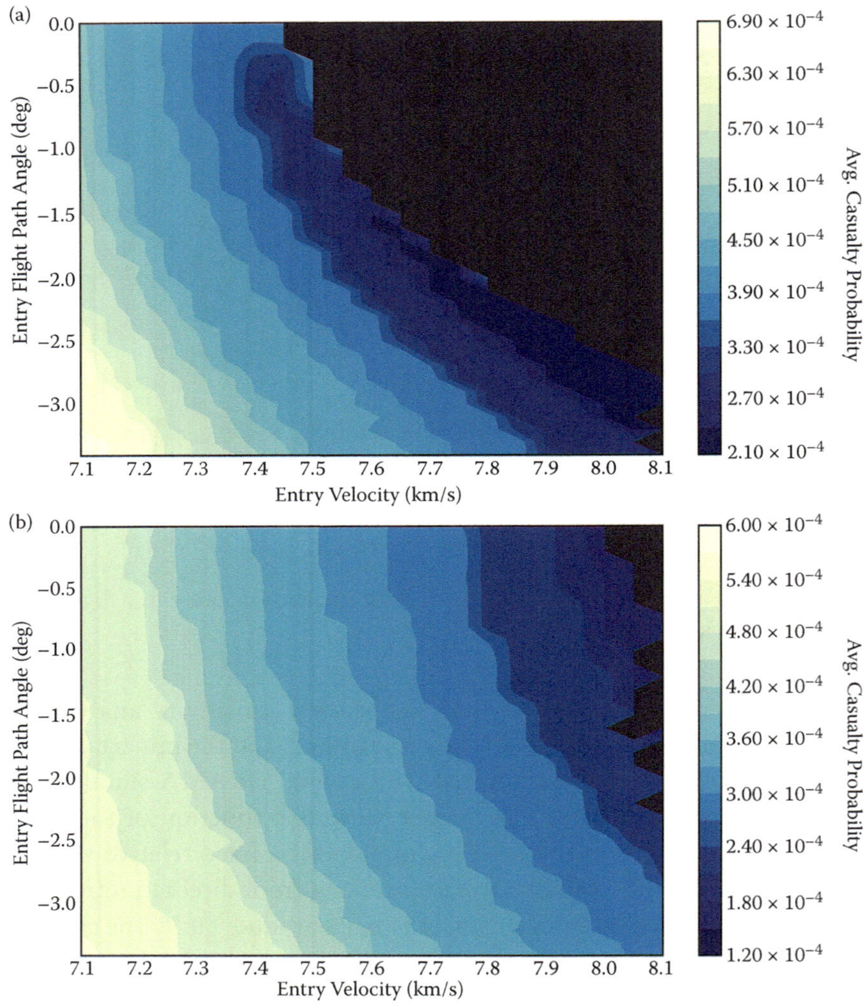

FIGURE 5.10 Casualty risk as function of entry velocity and flight path angle for (a) a 30-degree inclination orbit, and (b) a 120-degree inclination orbit.

orbit the demise of the spacecraft is greater for steeper reentries, whereas for the 120° orbit the demise is greater for shallower entries.

5.4 MITIGATION MEASURES

The space debris problem is nowadays internationally recognized, therefore mitigation measures are being taken and guidelines discussed. These can be divided into two classes: The avoidance or protection measures and the active and passive debris removal measures. The avoidance or protection measures include the design of satellites to withstand impacts by small debris, or the selection of safe procedures for operational spacecraft such as orbits with less debris, specific attitude configurations, or implementing active avoidance maneuvers to avoid collisions. On the other hand, measures for debris removal currently consist in limiting the creation of new debris (by prevention of in-orbit explosions and ensuring spacecraft subsystems reliability), implementing end-of-life disposal maneuvers to free some orbital protected regions, or to reenter in the atmosphere. Active debris removal is also being considered as a mean to stabilize the growth of space debris by removing from orbit some selected noncompliant objects. The e.Deorbit mission will target an ESA-owned derelict satellite in low orbit, capture it with a net or robotic arm technology, and reenter with a controlled atmospheric reentry (Biesbroek et al. 2014).

Acknowledging the fact that the projected growth in the number of satellites orbiting the Earth will increase in the future, space agencies and international organizations have been discussing and building a set of guidelines to ensure the sustainability of future space activities. The Inter-Agency Debris Coordination Committee (IADC) was founded in 1993 by ESA (Europe), NASA (the United States), the Japan Aerospace Exploration Agency (JAXA, Japan), and the Roscosmos Russian Federation. As of January 2017, the IADC also includes the Italian Space Agency (ASI, Italy), the Centre National d'Études Spatiales (CNES, France), the China National Space Administration (CNSA, China), the Canadian Space Agency (CSA, Canada), the German Aerospace Centre (DLR, Germany), the Korea Aerospace Research Institute (KARI, South Korea), the Indian Space Research Organisation (ISRO, India), the National Space Agency of Ukraine (NSAU, Ukraine), and the UK Space Agency (UKSA, United Kingdom). This international cooperation decided a set of space debris mitigation measures (Inter-Agency Space Debris Coordination Commitee, 2002), which includes:

1. Limitation of debris released during normal operations.
2. Minimization of the potential for on-orbit breakups (resulting from stored energy after the completion of mission operations, or during the operational phases of the mission and by avoiding intentional destruction and other harmful activities).

3. Post Mission Disposal in particular in geosynchronous regions and for objects passing through the LEO region.
4. Prevention of on-orbit collisions.

The IADC guidelines were presented to the United Nations Committee on the Peaceful Uses of Outer Space (UN COPUOS) and contributed to the creation of the Space Debris Mitigation Guidelines of the Committee on the Peaceful Uses of Outer Space to be considered for "the mission planning, design, manufacture and operational phases of spacecraft and launch vehicle orbital stages" (United Nations Office for Outer Space Affairs 2010):

1. Limit debris released during normal operations.
2. Minimize the potential for breakups during operational phases.
3. Limit the probability of accidental collision in orbit.
4. Avoid intentional destruction and other harmful activities.
5. Minimize potential for post-mission breakups resulting from stored energy.
6. Limit the long-term presence of spacecraft and launch vehicle orbital stages in the low Earth orbit region after the end of their mission.
7. Limit the long-term interference of spacecraft and launch vehicle orbital stages with the geosynchronous region after the end of their mission.

5.4.1 Mitigation Guidelines for Post Mission Disposal

In this section we focus on the third of the measures dictated by the IADC, namely Post Mission Disposal. A "25-year rule" was defined to limit the presence of satellites in the LEO region to no more than 25 years after their decommissioning. The 25-year limit was selected to ensure that a reasonable reduction in lifetime could be achieved without greatly affecting satellite resources. After 25 years a satellite has to be removed from the LEO protected region by placing it in a graveyard orbit or by disposing of it through atmospheric reentry. According to the IADC Space Debris Mitigation Guidelines (Inter-Agency Space Debris Coordination Commitee 2002) if "*a spacecraft or orbital stage is to be disposed of by re-entry into the atmosphere, debris that survives to reach the surface of the Earth should not pose an undue risk to people or property.*"

The low Earth orbit protected region (LEO region) is the spherical shell region that extends from the Earth's surface up to an altitude of 2000 km. The geosynchronous protected region (GEO region) is a segment of a spherical shell with a lower and upper altitude boundary of 200 km below and above the geostationary altitude of 35,786 km, and which is constrained by a latitude sector extending between plus and minus 15 degrees from south to north (Inter-Agency Space Debris Coordination Committee 2002; United Nations Office for Outer Space Affairs 2010).

At altitudes below 600 kilometers, spacecraft with a conventional area-to-mass ratio (i.e., conventional satellites have a value of area-to-mass ratio around 0.012 m^2/kg) will reenter within a few years due to atmospheric drag. Intervention to remove and prevent further creation of debris above that altitude should therefore be the primary focus of passive mitigation measures. As described in the document on the "Requirements on Space Debris Mitigation for ESA Projects" (ESA 2008) and the "ESA Space Debris Mitigation Compliance Verification Guidelines" (ESA 2015), end-of-life measures can be distinguished in: (1) Disposal, (2) passivation, and (3) reentry. Required measures for disposal currently cover spacecraft in LEO and GEO through a series of Operational Requirements (OR) (ESA 2008):

"OR-01. Space systems operating in the LEO protected region shall be disposed of by reentry into the Earth's atmosphere within 25 years after the end of the operational phase."

"OR-02. Space systems operating in the GEO protected region shall be disposed of by permanently removing them from the GEO protected region." The GEO disposal orbit should be almost circular (i.e., eccentricity less of equal to 0.005) and with a minimum perigee altitude above the geostationary altitude, which is given as a function of the solar radiation pressure coefficient of the space system at the beginning of its life and its cross-sectional area. This is done to take into account the eccentricity oscillation due to the effects of solar radiation pressure and to ensure that such oscillation would not make the orbit interfere with the GEO protected regions.

"OR-03. Where practicable and economically feasible, space systems outside the LEO and GEO protected regions shall implement means of end-of-life orbit disposal to avoid long-term interference with operational orbit regions, such as the Galileo orbit."

OR-04. Launcher stages shall also perform end-of-life disposal maneuvers by targeting "direct reentry as part of the launcher sequence." Alternatively, they should be injected into a LEO orbit with a maximum reentry time of 25 years. As other space systems, they should be removed from LEO and GEO protecting region and orbit that interfere with other operational orbits such as the one of the Galileo orbit.

OR-05. Passivation of the system (spacecraft or launcher stage) has to be completed within 2 months of the end-of mission.

End-of-life measures for reentry include:

OR-06. "For space systems that are disposed of by reentry," an "analysis has to be performed to determine the characteristics of fragments surviving to ground impact, and assess the total casualty risk to the population on ground assuming an uncontrolled reentry."

OR-07. Such a casualty risk has to be lower than 10^{-4} if an uncontrolled reentry is targeted; otherwise if the casualty risk is higher than the threshold of 10^{-4}, "a controlled reentry must be performed such that the impact footprint can be ensured over an ocean area, with sufficient clearance of landmasses and traffic routes."

The rate of compliance of missions to the end-of-life mitigation guidelines was analyzed by the ESA Space Debris Office in 2017). Between 2006 and 2015, the rate of compliance of LEO missions (including naturally compliant missions and satellites performing end-of-life maneuvers) was 53.3% for the payloads (corresponding to 60.3% of the payload mass), reaching end of life in the LEO protected region (Frey and Lemmens 2017). The compliant objects, with a lifetime after decommissioning of less than 25 years, include naturally compliant objects due to their initial altitude well inside the Earth's atmosphere (this constitutes the biggest part of the compliant share), compliant objects after a deorbit maneuver, or spacecraft having performed a maneuver leading to a direct reentry. In terms of mass, this share is constantly sloping downward. Between 2007 and 2016, 71.6% of the rocket bodies reaching end of life in the LEO protected region was compliant, and this fraction has remained virtually unchanged for 8 years in a row despite an increase in end-of-life maneuver activity.

5.4.2 Passive End-of-Life Disposal

In order to meet the mitigation guidelines LEO satellites at the end of their life would use the remaining propellant to perform either a perigee-lowering maneuver (to decrease the orbit perigee well inside the Earth's atmosphere to guarantee a reentry within 25 years) or a direct reentry. Spacecraft in GEO are instead currently re-orbited to quasi circular orbits outside the GEO protected ring, with a perigee line aligned with the Sun-Earth direction (where possible) in order to bind the long-term oscillations in the eccentricity caused by solar radiation pressure. Recently, ESA-funded projects on the design of disposal trajectories for *medium Earth orbits* (MEO) (Alessi et al. 2014; Rossi et al. 2015), *highly elliptical orbits* (HEO), and *libration Earth orbits* (LPO) (Armellin et al. 2014; Colombo et al. 2014; Colombo et al. 2015). These have demonstrated the possibility of exploiting natural orbit perturbations for designing passive mitigation strategies for debris disposal. Disposal strategies enhancing the effects of orbit perturbations have been further analyzed in LEO (Alessi et al. 2017), in MEO (Rosengren et al. 2015; Alessi et al. 2016; Armellin and San-Juan; Daquin et al. 2016; Gkolias et al. 2016), in GEO (Colombo and Gkolias 2017), and in HEO (Colombo et al. 2014; Armellin et al. 2015). Indeed, it was shown that, rather than performing an expensive maneuver to lower the perigee, the optimal maneuver should be given in a way to change the disposal orbit to another neighborhood orbit where the effect of orbit perturbations causes the orbit perigee to enter into the atmosphere. Indeed, the effects of luni-solar perturbation causes long-term oscillation on the eccentricity, which can be exploited so that the spacecraft's trajectory over a long period (from 5 to 70 years, depending on the initial orbit) could lead to natural reentry. This effect can be enhanced by solar radiation pressure, especially if considering a spacecraft equipped with large solar panels or a deployable reflective surface (Lücking et al. 2012, 2013). Moreover, resonances with the Earth's nonuniform potential can enhance the eccentricity growth effects.

5.4.2.1 An Example of End-of-Life Deorbiting Exploiting Luni-Solar Perturbations

One of the most beautiful demonstrations of how natural dynamics can be enhanced is given by the INTEGRAL mission designed by ESA, the United States, Russia, the Czech Republic, and Poland. The INTErnational Gamma-Ray Astrophysics Laboratory, launched in 2002, gathered some of the most energetic radiation from space (Eismont et al. 2003). A reentry of this spacecraft with a pure impulsive maneuver would have not been possible due to the limited amount of propellant left onboard. In an ESA-funded study, the end-of-life disposal of INTEGRAL mission—expected to end in 2016—was designed with a time window for disposal between January 1, 2013 and January 1, 2029. Reentry solutions with a delta-velocity requirement below 40–50 m/s were found (Colombo et al. 2014). The main perturbations acting on the dynamics of the reentry were luni-solar perturbations, which affect the evolution of eccentricity, inclination, and anomaly of the perigee measured with respect to the Earth-Moon plane. It was shown that depending on the set of initial elements, which depends on the date the reentry maneuver is performed, the proposed maneuver would then aim at further increasing or decreasing the eccentricity. In particular, if we focus on the natural evolution of the eccentricity under luni-solar perturbation and Earth's oblateness, when the nominal eccentricity is low, the optimal reentry maneuver further decrease the eccentricity value; as a consequence, the following long-term propagation will reach a higher eccentricity, corresponding to a reentry. In this case, the maneuver is more efficient (i.e., lower delta velocity is required) (Colombo et al. 2014). Once the initial disposal maneuver is performed, the spacecraft evolves under natural perturbations and the reentry can then be semicontrolled. The high inclination of HEOs represents an advantage as the final reentry phase can target regions at higher latitudes on the Earth's surface thereby reducing the ground hazard. In the case of HEOs, reentry is caused by luni-solar perturbation (not air drag), therefore the orbit reenter with quite a high eccentricity (high apogee and low perigee) and does not circularize. Due to the oscillations in eccentricity, the next optimal window for injecting the spacecraft into a reentry trajectory is between 2013 and the first half of 2018 for a final reentry in 2028. After that, the required maneuver would increase until reaching a next window for performing the maneuver between the second half of 2021 and the first half of 2026, for a reentry in 2028. These analytical studies were used for high fidelity parametric analyses performed by the ESA (Merz et al. 2015) to investigate the effect of a maneuver at apogee to change the perigee altitude. The final maneuver sequence was given at the beginning of 2015 and split into three major burns plus a touch-up for final fine-tuning. The spacecraft is now on its course to reentry in 2028 (see Figure 5.11).

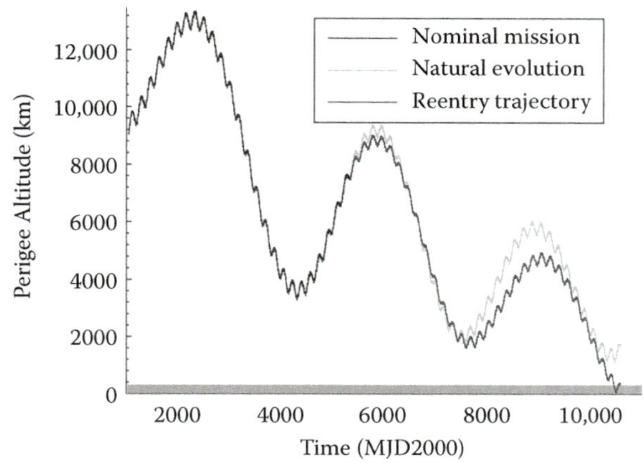

FIGURE 5.11 Reentry trajectory of the INTEGRAL spacecraft exploiting luni-solar perturbations. Black line: Natural trajectory; light gray line: continuation of the natural trajectory; medium gray line: deviated trajectory after the delta-velocity maneuver is performed. Such a maneuver will bring the spacecraft to enter the Earth's atmosphere represented by the gray band. Time in the x-axis is represented in Modified Julian Days since January 1, 2000 at 12:00 (MJD2000).

5.4.2.2 End-of-Life Deorbiting Exploiting Solar Radiation Pressure and Atmospheric Drag

Solar and drag sailing have also been proposed as passive end-of-life deorbiting methods, and technological demonstrators are under development. Drag sailing is of benefit for end-of-life disposal of small to medium satellites from orbits altitude up to 1000 km (Janovsky et al. 2003). Outside this altitude range, a region extending from high LEO (i.e., 1000 km) up to about 13,000 km can be identified where solar sailing is of interest (Lücking et al. 2012). In the drag-dominated regime the required area-to-mass ratio for a sail spacecraft to deorbit in a given time is primarily dependent on the semimajor axis, growing exponentially with increasing altitude. In the solar radiation pressure-dominated regime, the required area-to-mass ratio to deorbit in a given time strongly depends on both semimajor axis and inclination of the initial orbit. The deorbiting phase, at least in the first phase, is achieved on an elliptical orbit, not a circular orbit such as in the case of drag sail with inward deorbiting. The performance of the sailing strategy is determined by three parameters: (1) The required effective area-to-mass ratio to deorbit the spacecraft, which determines the sail size given the satellite's mass; (2) the time to deorbit; and (3) the augmented collision probability with the whole space debris population caused on and by the sail through its passage in the LEO protected region. During deorbiting the satellite crosses the "debris rings" of the Earth. The cumulative collision risk can be quantified as a function of the collisional cross section of the spacecraft in orbit and the time of exposure of this cross section to the flux of debris present in the environment. Current work is devoted to quantifying the interaction of the deorbiting strategy with the debris environment by computing the cumulative collision probability and assessing the global effect on the growth of the space debris population (Colombo et al. 2017) through one of the indices introduced in Section 5.3.1.1.3.

5.5 CONCLUSIONS

This chapter introduced the problem associated with the proliferation of space debris in orbit around the Earth. In particular, we focused on the analysis of the risk associated with space debris in orbit and on the ground and its quantification by the use of numerical indexes. Mitigation measures for limiting the increase of this risk and the sustainability of space activities are presented, with a particular focus on end-of-life disposal practices. We have introduced and given some examples of recent work on the exploitation of natural perturbations for enhancing the effects of end-of-life disposal maneuvers. The monitoring and mitigation of space debris, together with the monitoring and mitigation of potential hazardous asteroids (and the study of space weather) are part of the framework of space situational awareness.

ACKNOWLEDGMENTS

The authors acknowledge the use of the IRIDIS High Performance Computing Facility, and associated support services at the University of Southampton, in the completion of part of this work.

REFERENCES

Alby F., Alwes D., Anselmo L., Baccini H., Bonnal C., Crowther R., Tremayne-Smith R. 2004. The European space debris safety and mitigation standard. *Advances in Space Research*, 34(5), 1260–1263.

Alessi E. M., Rossi A., Valsecchi G. B., Anselmo L., Pardini C., Colombo C., ... Merz K. 2014. Effectiveness of GNSS disposal strategies. *Acta Astronautica*, 99, 292–302.

Alessi E. M., Deleflie F., Rosengren A. J., Rossi A., Valsecchi G. B., Daquin J., and Merz K. 2016. A numerical investigation on the eccentricity growth of GNSS disposal orbits. *Celestial Mechanics and Dynamical Astronomy*, 125(1), 71–90.

Alessi E. M., Schettino G., Rossi A., and Valsecchi G. B. 2017. Dynamical mapping of the LEO region for passive disposal design. *Paper Presented at the 68th International Astronautical Congress*, Adelaide, Australia.

Anselmo L. and Pardini C. 2005. Computational methods for reentry trajectories and risk assessment. *Advances in Space Research*, 35(7), 1343–1352.

Anselmo L. and Pardini C. 2015. Compliance of the Italian satellites in low earth orbit with the end-of-life disposal guidelines for space debris mitigation and ranking of their long-term criticality for the environment. *Acta Astronautica*, 114(Supplement C), 93–100.

Anselmo L. and Pardini C. 2016. Ranking upper stages in low earth orbit for active removal. *Acta Astronautica*, 122(Supplement C), 19–27.

Anselmo L., Rossi A., Pardini C., Cordelli A., and Jehn R. 2001. Effect of mitigation measures on the long-term evolution of the debris population. *Advances in Space Research*, 28(9), 1427–1436.

Armellin R. and San-Juan J. F. Optimal earth's reentry disposal of the Galileo constellation. *Advances in Space Research*. doi: https://doi.org/10.1016/j.asr.2017.11.028.

Armellin R., Di Mauro G., Rasotto M., Madakashira H. K., Lara M., and San-Juan J. F. 2014. *End-of-Life Disposal Concepts for Lagrange-Points and HEO Missions*. Final Report ESA contract AO/1-7210/12/F/MOS.

Armellin R., San-Juan J. F., and Lara M. 2015. End-of-life disposal of high elliptical orbit missions: The case of INTEGRAL. *Advances in Space Research*, 56(3), 479–493.

Bastida Virgili B. and Krag H. 2013. Active debris removal for LEO missions. *Paper Presented at the Sixth European Conference on Space Debris*, Darmstadt, Germany.

Biesbroek R., Hüsing J., and Wolahan A. May 6, 2014. System and concurrent engineering for the e.Deorbit mission assessment studies. *Paper Presented at the e.Deorbit Symposium—ESA, Conference Centre Leeuwenhorst*, Amsterdam.

BIO Intelligence Service 2014. Environmental Impact Assessment Analysis: Technical Note D2.3.1: Environmental impact assessment of ESA projects – Sentinel 3. In E. S. A. s. contract (Ed.).

Colombo C. and Gkolias I. April 18–21, 2017. Analysis of the orbit stability in the Geosyncronous region for end-of-life disposal. *Paper Presented at the 7th European Conference on Space Debris*, ESA/ESOC, Darmstadt/Germany.

Colombo C., Letizia F., Alessi E. M., and Landgraf M. January 26–30, 2014. End-of-life Earth reentry for highly elliptical orbits: The INTEGRAL mission. *Paper Presented at the 24th AAS/AIAA Space Flight Mechanics Meeting*, Santa Fe, New Mexico.

Colombo C., Letizia F., Soldini S., Lewis H., Alessi E. M., Rossi A., Vasile M., Vetrisano M., and van der Weg W. 2014. *End-of-Life Disposal Concepts for Lagrange-Point and Highly Elliptical Orbit Missions*. Final Report, ESA/ESOC contract No. 4000107624/13/F/MOS.

Colombo C., Alessi E. M., van der Weg W., Soldini S., Letizia F., Vetrisano M., Vasile M., Rossi A., and Landgraf M. 2015. End-of-life disposal concepts for libration point orbit and highly elliptical orbit missions. *Acta Astronautica*, 110, 298–312.

Colombo C., Rossi A., Dalla Vedova F., Braun V., Bastida Virgili B., and Krag H. 2017. Drag and solar sail deorbiting: Reentry time versus cumulative collision probability. *Paper Presented at the 68th International Astronautical Congress 2017*, Adelaide, Australia.

Daquin J., Rosengren A. J., Alessi E. M., Deleflie F., Valsecchi G. B., and Rossi A. 2016. The dynamical structure of the MEO region: Long-term stability, chaos, and transport. *Celestial Mechanics and Dynamical Astronomy*, 124(4), 335–366.

Dolado-Perez J. C., Revelin B., and Di-Costanzo R. 2015. Sensitivity analysis of the long-term evolution of the space debris population in LEO. *Journal of Space Safety Engineering*, 2(1), 12–22.

Durrieu S. and Nelson R. F. 2013. Earth observation from space—The issue of environmental sustainability. *Space Policy*, 29(4), 238–250.

Eismont N. A., Ditrikh A. V., Janin G., Karrask V. K., Clausen K., Medvedchikov A. I., ... Yakushin N. I. 2003. Orbit design for launching INTEGRAL on the proton/block-DM launcher. *Astronomy and Astrophysics*, 411(1), L37–L41.

ESA Space Debris Office, 2017. ESA's Annual Space Environment Report. Produced with the DISCOS Database. https://www.sdo.esoc.esa.int/environment_report

ESA, 2008. Requirements on Space Debris Mitigation for ESA Projects.

ESA, 2015. Space Debris Mitigation Compliance Verification Guidelines.

Flohrer T., Krag H., and Klinkrad H. 2009. ESA's process for the identification and assessment of high-risk conjunction events. *Advances in Space Research*, 44(3), 355–363.

Flury W. 1995. The space debris environment of the earth. *Earth, Moon, and Planets*, 70(1), 79–91.

Frey S. and Lemmens S. 2017. Status of the space environment: Current level of adherence to the space debris mitigation policy. *Journal of the British Interplanetary Society*, 70(2–4), 118–124.

Gelhaus J., Sanchez-Ortiz N., Braun V., Kebschull C., de Oliveira J. C., Dominguez-Gonzalez R., ... Vorsmann P. April 22–25, 2013. Upgrade of DRAMA-ESA's space debris mitigation analysis tool suite. *Paper Presented at the 6th European Conference on Space Debris*, ESA/ESOC, Darmstadt, Germany.

Gkolias I., Daquin J., Gachet F., and Rosengren A. J. 2016. From order to chaos in earth satellite orbits. *The Astronomical Journal*, 152(5), 119.

Inter-Agency Space Debris Coordination Commitee, 2002. IADC Space Debris Mitigation Guidelines (Revision 1–2007 ed.).

International Standards Organisation, 2011. Space Systems—Space Debris Mitigation.

Janovsky R. et al. 2003. End-of-life de-orbiting strategies for satellites. *Paper Presented at the 54th International Astronautical Congress*, Bremen, Germany.

Johnson N. L., Krisko P. H., Liou J. C., and Anz-Meador P. D. 2001. NASA's new breakup model of evolve 4.0. *Advances in Space Research*, 28(9), 1377–1384.

Kebschull C., Radtke J., and Krag H. September 29 to October 3, 2014. Deriving a priority list based on the environmental criticality. *Paper Presented at the 65th International Astronautical Congress*, Toronto, Canada.

Klinkrad H., Fritsche B., Lips T., and Koppenwallner G. 2006. Reentry prediction and on-ground risk Estimation. In S. P. Books (Ed.), *Space Debris*. Berlin, Heidelberg: Springer.

Koppenwallner G., Fritsche B., Lips T., and Klinkrad H. November 8–11, 2004. Scarab-a multi-disciplinary code for destruction analysis of space-craft during reentry. *Paper Presented at the Fifth European Symposium on Aerothermodynamics for Space Vehicles*, Cologne, Germany.

Krag H., Serrano M., Braun V., Kuchynka P., Catania M., Siminski J., … McKissock D. 2017. A 1 cm space debris impact onto the sentinel-1A solar array. *Acta Astronautica*, 137(Supplement C), 434–443.

Krisko P. H. 2011. Proper implementation of the 1998 NASA breakup model. *Orbital Debris Quarterly News*, 15(4), 1–10.

Letizia F., Colombo C., and Lewis H. G. 2015. Analytical model for the propagation of small debris objects clouds after fragmentations. *Journal of Guidance, Control, and Dynamics*, 38(8), 1478–1491.

Letizia F., Colombo C., Lewis H. G., and Krag H. 2016. Assessment of breakup severity on operational satellites. *Advances in Space Research*, 58(7), 1255–1274.

Letizia F., Colombo C., Lewis H. G., and Krag H. April 18–21, 2017. Extending the ECOB space debris index with fragmentation risk estimation. *Paper Presented at the 7th European Conference on Space Debris*, ESA/ESOC, Darmstadt, Germany.

Lewis H. G., George S. G., Schwarz B. S., and Stokes H. 2013. Space Debris Environment Impact Rating System. *Paper Presented at the Sixth European Conference on Space Debris*, Darmstadt, Germany.

Liou J. C. and Johnson N. L. 2009. A sensitivity study of the effectiveness of active debris removal in LEO. *Acta Astronautica*, 64(2), 236–243.

Lips T. and Fritsche B. 2005. A comparison of commonly used reentry analysis tools. *Acta Astronautica*, 57(2), 312–323.

Lücking C., Colombo C., and McInnes C. R. 2012. A passive satellite deorbiting strategy for medium earth orbit using solar radiation pressure and the J_2 effect. *Acta Astronautica*, 77, 197–206.

Lücking C., Colombo C., and McInnes C. R. 2013. Solar radiation pressure-augmented deorbiting: Passive end-of-life disposal from high altitude orbits. *Journal of Spacecraft and Rockets*, 50(6), 1256–1267.

Martin C., Cheese J., Brandmueller C., Bunte K., Fritsche B., Lips T., … Sanchez N. A. 2005. A debris risk assessment tool supporting mitigation guidelines. *Paper Presented at the 4th European Conference on Space Debris*, ESA/ESOC, Darmstadt, Germany.

MASTER, 2013. Meteoroid and Space Debris Terrestrial Environment Reference (MASTER). https://sdup.esoc.esa.int. Retrieved 01/10/2017.

McKnight D. S. 2016. Orbital debris hazard insights from spacecraft anomalies studies. *Acta Astronautica*, 126 (Supplement C), 27–34.

Merz K., Krag H., Lemmens S., Funke Q., Böttger S., Sieg D., … Southworth R. October 19–23, 2015. Orbit aspects of end-of-life disposal from highly eccentric orbits. *Paper Presented at the 25th International Symposium on Space Flight Dynamics ISSFD*, Munich, Germany.

NASA Orbital Debris Quarterly News, 2016. 20(1&2).

Pardini C. and Anselmo L. 2009. Assessment of the consequences of the fengyun-1C breakup in low earth orbit. *Advances in Space Research*, 44(5), 545–557.

Pardini C. and Anselmo L. 2017. Revisiting the collision risk with cataloged objects for the iridium and COSMO-SkyMed satellite constellations. *Acta Astronautica*, 134(Supplement C), 23–32.

Patera R. P. 2008. Hazard analysis for uncontrolled space vehicle reentry. *Journal of Spacecraft and Rockets*, 45(5), 1031–1041.

Rosengren A. J., Alessi E. M., Rossi A., and Valsecchi G. B. 2015. Chaos in navigation satellite orbits caused by the perturbed motion of the moon. *Mon. Not. R. Astron. Soc.* 449, 3522–3526.

Rossi A. et al. 2015. *Disposal Strategies Analysis for MEO Orbits* (ESA/ESOC, Trans.) (Vol. Final Report).

Rossi A., Cordelli A., Farinella P., Anselmo L., and Pardini C. 1997. Long term evolution of the space debris population. *Advances in Space Research*, 19 (2), 331–340.

Rossi A., Lewis H., White A., Anselmo L., Pardini C., Krag H., and Bastida Virgili, B. 2016. Analysis of the consequences of fragmentations in low and geostationary orbits. *Advances in Space Research*, 57(8), 1652–1663.

Rossi A., Valsecchi G. B., and Alessi E. M. 2015. The criticality of spacecraft index. *Advances in Space Research*, 56(3), 449–460.

Sdunnus H., Beltrami P., Klinkrad H., Matney M., Nazarenko A., and Wegener P. 2004. Comparison of debris flux models. *Advances in Space Research*, 34(5), 1000–1005.

Symonds K., Fornarelli D., Mardle N., Ormston T., Flohrer T., and Marc X. May 5–9, 2014. Operational reality of collision avoidance maneuver execution. *Paper Presented at the SpaceOps*, Pasadena, CA.

Trisolini M., Lewis H. G., and Colombo C. 2015. Survivability and demise criteria for sustainable spacecraft design. *Paper Presented at the 66th International Astronautical Congress*, Jerusalem.

Trisolini M., Lewis H. G., and Colombo C. October 2016. Spacecraft design optimisation for demise and survivability. *Paper Presented at the 67th International Astronautical Congress*, Guadalajara, Mexico.

U.S. Government, 1997. U.S. Government Orbital Debris Mitigation Standard Practices.

United Nations Office for Outer Space Affairs, 2010. Space Debris Mitigation Guidelines of the Commetee on the Peacefull Uses of Outer Space (Vol. 9). Vienna, United Nations.

Utzmann J., Oswald M., Stabroth S., Voigt P., and Retat I. October 1–5, 2012. Ranking and characterization of heavy debris for active removal. *Paper Presented at the 63rd International Astronautical Congress*, Naples, Italy.

Vallado D. A. and McClain W. D. 2007. Chapter 8: Special perturbation techniques. In S. T. Library (Ed.), *Fundamentals of Astrodynamics and Applications* (Fourth Edition ed., pp. 551–573). New York: Microcosm Press.

Wiedemann C., Gamper E., Horstmann A., Braun V., and Stoll E. April 18–21, 2017. The contribution of NaK Droplets to the space debris environment. *Paper Presented at the 7th European Conference on Space Debris*, Darmstadt, Germany.

Yasaka T. 2011. Can we have an end to the debris issue? *Paper Presented at the 62nd International Astronautical Congress*, Cape Town, South Africa.

CHAPTER 6

Commercial Space Risks, Spacecraft Insurance, and the Fragile Frontier

Mark Williamson

CONTENTS

6.1 INTRODUCTION

The inherent temperature extremes and radiation hazards of the natural space environment, coupled with the possibility of physical impact from both natural and manmade debris objects, make space a uniquely difficult place to do business. As such, the space environment presents a challenge to designers and operators of our space technology assets, from commercial communications, navigation, and weather satellites to orbiting space stations and spacecraft on missions beyond Earth orbit.

Although parts of the Earth's orbital environment have been used by private companies since the mid-1960s, the inherent risks and attendant liability regime associated with spaceflight have received increased attention in recent years with the advent of commercial supply missions to the International Space Station, the increasing use of compact satellites known as "cubesats," and a potential market for suborbital "space tourism" flights. Although government/space agency–funded missions—from navigation and surveillance satellites to crewed capsules and space stations—face most of the same risks, the loss of a commercial space asset could result in the failure of an enterprise or, potentially, the demise of an entire space application or industry segment.

This contribution provides an overview of the risks encountered by space-based assets and the engineering solutions provided by space technology. For immediate commercial risks, it explains the mitigating role of the space insurance market, the coverage available, and other factors such as market capacity and premium rates for space risks. With regard to the long-term sustainability of space-based assets, it highlights the fragility of orbital and planetary environments and the challenge of protecting them for study and use by future generations.

6.2 COMMERCIAL SPACE RISKS

The challenges to and risks encountered by technology designed for use here on Earth are reasonably well understood in that even nonexperts understand that aspects of the natural environment have the potential to cause damage. For example, the Sun's heat can melt materials, its ultraviolet rays may discolor certain plastics and its overall intensity can cause colors to fade. As for more extreme environments, it is common knowledge that aircraft flying at altitude experience low air pressure and that submarines at depth experience extremely high water pressure.

The environmental risks to space-based technology are, arguably, less widely known because it is far more difficult to experience them, either personally or vicariously. That said, many texts have been published on the engineering design of space-based systems and the subject of space technology is well documented (Williamson 1990, 2001).

In essence, the space environment is an alien environment, which is inherently hostile to anything born or made on Earth. For example, while the sun provides warmth and light and generates solar power, various parts of its spectrum are damaging to materials—much as on Earth, but even more so due to the lack of a protective atmosphere. Moreover, the distribution of heat in space is reliant only on radiation, there being no convection in an airless environment and only very local conduction. As a result, the sunlit side of a spacecraft may be some 200°C above zero while its shadowed side is some 200°C below (dependent on specific conditions); this presents a significant challenge for spacecraft thermal engineers. Likewise, excess heat (from electronic equipment, for instance) can only be removed from a spacecraft by radiation, which places constraints on the power used by equipment and drives requirements for the surface area available for radiators. Although thermal engineering is relevant to terrestrial devices, it is crucial to the design of space-based systems.

Of course, the thermal environment is but one challenge faced by space-based assets; gravitational effects, solar radiation pressure, Galactic radiation and particles, space weather, and space debris offer their own particular challenges, alongside a gamut of human-related risks from signal jamming to anti-satellite weaponry. We should, moreover, recognize the ever-present risk of human error at every stage of a mission from design, through manufacturing and testing, to launch and in-orbit operation; sometimes equipment fails, people make mistakes, and missions are compromised.

Needless to say, the addition of a crew or passengers to a space-based system requires quite another level of engineering design and risk mitigation measures. However, the remit of this contribution is limited to the risks encountered by commercial space-based assets, predominantly represented by various types of communications satellite but increasingly joined by other satellite applications and by commercial space station delivery systems.

The key point is that risks are inevitable and need to be managed. Two types of risk management or risk mitigation are considered here: technical and financial. The first relates to the resiliency and reliability aspects of the engineering design process and the second relates to financial risk mitigation in the form of space insurance.

6.3 ENGINEERING RISK MITIGATION

Risk mitigation—the reduction of the probability of failure—is a central tenet of satellite manufacturing, which has evolved since the early days of spacecraft development. The main reason it is embedded at every stage from initial design to final testing relates to the working environment of the spacecraft as a product: to a greater extent than any other commercial product the spacecraft is inaccessible once launched, and thus it cannot be maintained or repaired in the same way as most Earth-based products. Another reason is that commercial spacecraft manufactured without this preoccupation with quality and reliability would be unlikely to secure insurance or project financing.

When introduced to a new satellite or system, one of the first questions asked by the insurance community is "What is the heritage?" In other words, to what extent have the spacecraft design and its constituent components been tested and operated in the space environment. In recognition of the potential risks, engineers and insurers alike tend to be conservative; this means that most spacecraft designs are based on previously proven designs and space-qualified components. However, new technology must be introduced if a field is to progress and this is achieved, in parallel with risk reduction, by establishing an integrated quality control system.

There are many aspects to a quality control system implemented for spacecraft design, manufacturing, and testing, but its bedrock lies in the experience of prime and subcontractors and the skills of a highly trained workforce. Historically, spacecraft have been built under labor-intensive procedures, with continual checking and quality control, and using proven materials and components. Testing at every stage—component, assembly, system…all the way to fully integrated spacecraft—has been part of the ethos of satellite manufacturing. As a result, it is not uncommon for a commercial communications satellite project to run for 3 years, from contract signature to launch, and major science missions for a decade or more.

While this is still the case for the larger more expensive spacecraft, in recent years there have been moves toward mass-production, typically for the smaller constellation spacecraft, and the industry appears to be on the verge of a revolution—as evidenced by the intention of OneWeb to build fifteen 150 kg-class satellites a week for its high-speed Internet constellation (Williamson 2016). Although, in general, mass production can increase reliability through repeatability ("practice makes perfect"), the reduction in testing inherent in such satellite manufacturing developments creates an issue for insurers. The general rule—that only the first few satellites are subjected to a full range of tests, with subsequent models treated to an extent as "carbon copies"—means that some defects may not become evident until several (perhaps many) satellites have been launched.

Although the detailed pros and cons of any new production schemes are beyond the scope of this chapter, it remains the case that if they are to be successful they will have to adhere to the common practice of "engineering-in" quality and reliability. If they do not, the projects will be tantamount to launching space junk!

One of the key aspects of a reliable spacecraft design is the recognition of the possibility of failure and the consequent incorporation of redundant components. As a result, most spacecraft subsystems and payloads incorporate "spare" hardware, which can be substituted (typically automatically) in case the primary system fails. For example, thrusters are usually mounted in redundant pairs, sensors are mutually redundant, and backup control systems are the norm. Larger, more massive items may be fitted as a three-for-two redundant system (in which two are required and the third is a spare for either of the pair); likewise, reaction or momentum wheels (for pointing control) are typically supplied in a four-for-three system. Amplifiers in a communications payload may instead use a ring-redundancy system, whereby several mutually redundant amplifiers are arranged in an interconnected "ring" (e.g., 10-for-8 redundancy, with two spares for every eight required).

In some cases, a balance between specifying heritage parts and introducing new technology is achieved through redundancy techniques, for example by installing a prime unit of new design and a redundant unit with a good heritage; if the new unit fails, the heritage backup is available.

Of course, in some cases redundant units cannot be provided, because of lack of space or because of limitations in the mass budget, but in these cases redundancy is often provided by the subcomponents. For example, it would not be mass-efficient to duplicate an entire propulsion system just in case part of it fails, but individual valves, thrusters, and ignition control systems would be provided to make the system more tolerant of failure. The key here is to reduce to a minimum any single points which may cause the failure of an entire system.

For all systems, whether or not redundancy is incorporated, the driving design requirement is one of risk mitigation. Unfortunately, however, components and subsystems still fail, occasionally as a result of design errors, sometimes due to manufacturing defects and also due to deficiencies in testing. At this point, the system's owner or operator may have to resort to the insurance policy, which was, hopefully, purchased as a backup risk mitigation measure.

6.4 SPACE INSURANCE

6.4.1 Overview

A key part of the financial solution for a commercial space program—beyond funding mechanisms for developing, building, and operating the system—is space insurance, which acts in a similar way to motor insurance, property insurance, or any other terrestrial application. In essence, an amount of money (the premium) is paid to insurance underwriters to cover a specified asset value under various loss scenarios; if a loss occurs, a claim is made on the policy and an amount of money is paid to the policyholder.

As with any insurance contract, the guiding principle is that such a payment should not place the claimant in a financially improved situation compared with that prior to payment of any claim. But again, as with any other insurance, the advantage of being insured is that property can be replaced and businesses can continue to operate in a way which would be impossible without insurance. For many enterprises, insurance is also a prerequisite for bank loans and the like.

The steps in arranging a space insurance policy will be familiar to anyone with a vehicle or domestic policy: one must research insurers or engage a broker to make recommendations; provide relevant information; consider coverage and excess values; consider extras and options; compare quotes; pay the insurance premium; and claim on the policy if necessary.

The space insurance market has a number of actors and relationships, as follows (see also Figure 6.1). The client is typically a satellite owner or operator who wishes to insure anything from a single spacecraft to an entire system (including both space and ground segments). The client engages a satellite manufacturer (otherwise known as a prime contractor), which manages a team of subcontractors to design, build, and test a product or system. The client also engages a launch supplier to place the satellite, or satellites, in orbit (although, contractually, this can be managed by the prime contractor under a so-called delivery-in-orbit contract). The client also engages a team of technical consultants, lawyers, and an insurance broker. Among other things, the broker ascertains the client's requirements and designs an insurance policy (which generally includes a mathematical loss formula) to cover the specified risks; the broker then negotiates with insurers or "underwriters" who agree to "underwrite" the policy (specifically, they enter a legally binding contractual agreement to pay valid claims which occur during the period of insurance cover).

Although the market is dynamic and market shares are constantly in flux, the global space insurance market is currently brokered by three leading brokerages (with a handful of smaller ones representing particular clients). Likewise, at any given time, some five or six lead underwriters in space

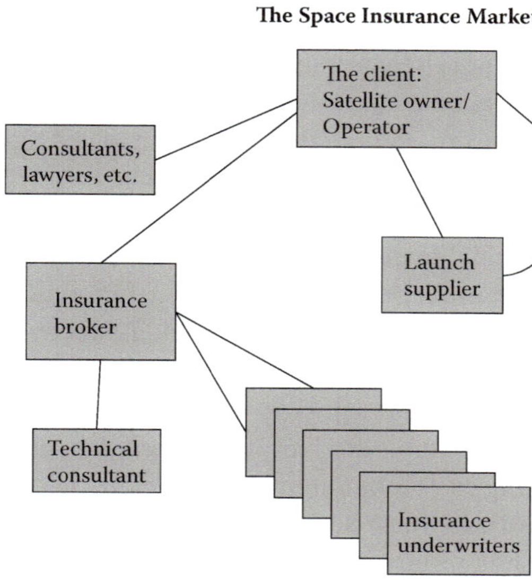

FIGURE 6.1 Key actors and relationships in the space insurance market, as described in Section 6.4.1.

can be identified in a global market of some 40 insurance companies. The amount of capital assigned by underwriters to space insurance, otherwise known as market capacity, varies but is currently around $750 million (see Section 6.4.5).

Although some leading space underwriters are based in France, Germany, and the United States, by far the largest concentration resides in London, mostly at Lloyd's of London. In the general scheme of potential space-related risks, Lloyd's has identified, developed, and adopted a number of Realistic Disaster Scenarios (RDS) and reserved funds for specified "extreme space events" (Gubby et al. 2014). The RDS report lists four potential risks as follows (Lloyd's 2017):

1. An anomalously large solar energetic particle event affecting many satellites (frequency considered to be 1-in-100 years);

2. A generic defect causing undue space weather sensitivity in a class or classes of satellites (1-in-50 years);

3. A generic defect causing unforeseen failures in a class or classes of satellites (1-in-20 years);

4. Collision with orbiting space debris (1-in-15 years).

The detailed mechanisms involved in solar events, space weather, and orbital debris are covered in detail elsewhere in this volume. It is clear from the RDS that the space insurance community is taking these potential threats seriously, but it is important to note that the risk involved in each of these scenarios is not considered equal; indeed far from it.

Although solar events have been shown to affect satellites and have the potential to cause large financial losses, a number of mitigating factors are already "built-in." For example, it has been known for some time that, as the scale of electronic components has decreased, microprocessor "chips" have become more susceptible to energetic particle events. Indeed, although satellite equipment suffers regularly from particle-induced charging and subsequent electrostatic discharge (ESD) events, they are rarely unrecoverable. Traveling wave tube amplifiers (the main high-power amplifiers utilized on board satellites) suffer spontaneous switch-offs, but they incorporate automatic restart circuits. More sensitive electronic hardware incorporates radiation shielding as part of the package design and is increasingly "hardened" against radiation in the first place (a technique pioneered in military circles). Nevertheless, the continued increase in chip-component density is causing concern, among space agencies and insurers alike, as nonrecoverable "single event effects" due to high-energy electrons are becoming more prominent.

As for "space weather" as a specific cause of insurance claims, one leading space underwriter (at Atrium) attributes 19% of satellite anomalies "at least in part" to space weather, but points out that the term "anomaly" can cover trivial events such as a "spurious switch off" (cured by a simple reset, which the end-user of the satellite may not even notice). Thus, in the past quarter century insurance claims due to space weather have amounted to $275 m, just 2% of total claims in that period (Wade 2013). As a result, the market generally considers it in terms of a "watching brief" only; that said, if a space weather event results in a large loss the situation could change, literally overnight. Likewise, in relation to the debris threat, the insurance market tends to be reactive and is seemingly content to "wait for the big event."

Based on its historical experience, the space insurance market focuses predominantly on hardware defects, anomalies, and failures—whether isolated or generic—caused by factors such as design deficiencies, human/manufacturing errors, insufficient testing, and so on. Although not always easy to prove or understand, this type of risk is considered palpable within the insurance industry, arguably because it can be related to terrestrial experience. In the same way that most mortals will never get to experience the space environment, insurers—along with everyone else—find the risks presented by the natural space environment more difficult to assess.

Indeed, when insurers consider the risk of failure by analyzing actual losses at different points in a mission they generally apportion 37% of the risk to the launch (covering anything from rocket explosions to stage separation and guidance errors) and 63% to the satellite; notably, some 34% of satellite losses occur in the first 2 months in orbit due to solar array deployment issues, propulsion problems, component infant mortality, and so on (Kunstadter 2016).

A key reason behind the introduction of the RDS was the potential risk of generic defects in components resulting from a common industrial practice—consolidation of the supply chain. The market for space-qualified components has always been small and key components have historically been produced by only two or three specialized contractors, but a cocktail of corporate competitive forces, financial market pressures, and the general march of globalization has further reduced choice in component supply. This increases the risk that an undetected generic defect or weakness in a basic component part—such as a momentum wheel or traveling wave tube amplifier—will affect not only spacecraft from a given prime contractor, but also those from different prime contractors. If this part was particularly sensitive to a given type or strength of space weather event, for example, such an event might affect a disproportionately large percentage of the in-orbit satellite fleet. The argument for space insurance is clear.

6.4.2 History

Despite the low public profile of space insurance, it has been under development as a business for more than half a century. The first space-related insurance policy was written for the Intelsat I communications satellite (colloquially known as "Early Bird"), which was launched in April 1965. The policy covered only third party liability and prelaunch risks (see Section 6.4.4), thus reflecting the insurance industry's limited engagement with the new technology; coverage for the launch itself and the ensuing in-orbit lifetime was not provided. Moreover, the policy was placed with the aviation insurance community, because the technology in that field was closer to space technology than any other.

Of the first six launches for the Intelsat consortium (Intelsat I, four Intelsat IIs, and an Intelsat III) only four successfully reached their designated position in geostationary orbit (GEO). This led Comsat, the operator of the Intelsat system, to seek launch insurance for the next seven Intelsat III spacecraft (designated F2–F8). They succeeded in doing so, but because insurers of the time were averse to the unfamiliar risks involved, the policy was written on the basis of "one failure deductible" in each of two series of launches, meaning that the first failure in each series would not be covered. Unfortunately for Comsat there was a launch failure in the first series (F5) and a satellite propulsion malfunction in the second (F8's apogee kick motor failed to place the satellite in GEO, but it later achieved orbit using station-keeping propellant at the expense of a reduced in-orbit lifetime). The unrelated hardware failures meant that the insurance industry did not have to pay any claims and Comsat was effectively "out of pocket."

Again, Comsat wanted to broaden the scope of insurance cover by including orbital positioning in the definition of launch insurance and such a policy was negotiated for the Intelsat IV series of the early 1970s (Hill 1983). However, policies were still being issued on a one-failure-deductible basis and the failure of one of the first eight Intelsat IV spacecraft (F6) to reach orbit, once again, did not result in an insurance payout.

By the mid-1970s, the reliability of the main U.S. launch vehicles (Atlas and Delta) was improving sufficiently for the insurance industry to agree (in 1975) to waive the deductible for the first time. The first satellite to benefit was the Marisat maritime communications satellite launched in 1976. Another insurance innovation initiated in 1975 was coverage for in-orbit failures, which paved the way for the comprehensive insurance of space risks and provided the basic template for today's satellite insurance policies. The first policy to cover a satellite's orbital lifetime was negotiated by RCA for its Satcom series (Hill 1983); it was based on 10 years experience of geostationary satellites, which had given the insurance community sufficient encouragement to provide cover for the operational phase and allowed a sensible partial failure clause to be written (see Sections 6.4.4 and 6.4.6).

6.4.3 Market

Today's space insurance policies provide coverage for a range of predominantly commercial, as opposed to government-owned, satellites based in a variety of orbits. Government entities, such as space agencies, and military/defense-related organizations tend not to engage with the space insurance market (though there are exceptions), at least in part because they find it difficult to justify the additional expense. For example, if a space telescope fails in orbit, there are no business repercussions and, ostensibly, only scientific data collection is impacted. Military systems are difficult to insure because of national security and technology transfer issues. Exceptions for a space agency mission might include insurance on a launch-only basis, which would provide funds for a replacement spacecraft to be built; and insurance for a military communications satellite could include insuring "service deliverables," such as geographical coverage and data rates, and/or contract incentive payments.

However, beyond launch coverage, the great majority of space insurance is currently attached to communications satellites in geostationary orbit and constellations of satellites in medium and low Earth orbits (MEO and LEO). A smaller percentage covers imaging satellites, typically based in a type of polar orbit known as heliosynchronous orbit, and (to a very limited degree) weather and navigation satellites in LEO, MEO, and GEO. As alluded to earlier, in recent years the market for commercial cargo deliveries to the International Space Station has developed, along with appropriate insurance, but this is still a small part of the overall insurance market (in 2016, e.g., communications satellites accounted for some 76% of insurers' income, imaging satellites around 19.5%, and ISS cargo and other satellites together only about 4.5%).

It is instructive to examine the limited size of the market. According to statistics compiled in late 2016, there were 471 active satellites in geostationary orbit, of which 214 were insured (i.e., just 45%). Likewise, active satellites in nongeostationary orbits (mainly LEO) totaled 1077, 58 (or 5%) of which were insured (Kunstadter 2016). The figures show that satellite insurance is still a somewhat exclusive market.

As for launch statistics, in 2016 there were 85 launches to orbit, two launch failures (2.4%) and, depending on definitions, four "anomalous" launches (e.g., where engines shut down early or satellites were left in a lower-than-intended orbit). According to Seradata (Todd 2016), the "overall launcher related failure rate since 1990" is 6%, but the chance of a failure in the first five flights of a new vehicle is 20%, which indicates that launching rockets remains a risky business and explains the need for insurance. In 2016, 35 launches were insured (41%) and there was one insured launch failure (and therefore one claim). The recent rise in the number of very small (10–50 kg) satellites and 1–10 kg cubesats has made it necessary not to take raw numbers at face value: the 85 launches mentioned deployed 217 satellites, 109 of which were cubesats (50%); 48 of the 108 non-cubesat satellites were insured (44%).

6.4.4 Coverage

The coverage offered by the majority of space insurance policies addresses the four main phases of a satellite's existence, thus providing manufacturing, prelaunch, launch, and in-orbit coverage.

Manufacturing coverage addresses the manufacturing, assembly, and testing of hardware at the appropriate industrial facilities: it includes such standard categories as production line, material damage, and work-in-progress coverage and manufacturers' all risks coverage. It is similar to nonspace industrial coverage and is therefore not placed in the space market.

Prelaunch coverage addresses factors such as transportation to the launch site, temporary storage, integration with the launch vehicle, prelaunch testing, and propellant loading. Coverage terminates at contractually specified times: intentional ignition of a launch vehicle's first stage engines, release of the hold-down clamps, or the moment of liftoff.

Launch coverage begins as prelaunch coverage ends and addresses factors such as delivery from the launch pad to geostationary transfer orbit (GTO), separation of the satellite from the launch vehicle, transfer to geostationary orbit, drift orbit maneuvers, solar array and antenna deployments, and in-orbit testing and commissioning. Launch policies also cover an initial period of operation in orbit (typically one year, which would include at least one full eclipse period when the Sun is eclipsed by the Earth and the satellite is subject to a more dynamic thermal environment and has to rely on its batteries).

In-orbit coverage addresses issues that may arise during the operational lifetime of the satellite and is generally renewed on an annual basis. Following first renewal, after the launch-plus-one year policy has expired, it typically covers the *total loss* of the satellite or the *partial loss* of the satellite's capability. Partial loss is based mainly on propellant deficiencies (which reduce station-keeping lifetime), power subsystem failures (which reduce the number of operable transponders and thus revenue-earning capability) and/or outright transponder failure (a transponder is an amplifier chain in the communications payload).

In addition to policies covering the four main phases—manufacturing, prelaunch, launch, and in-orbit—a number of other policies can be arranged. For example:

- *Manufacturers' incentives and penalties*: To cover either additional payments for meeting prearranged deadlines or similar, or charges for failing to do so, poor performance, etc.

- *Service interruption coverage*: To cover nonrefundable production and transmission costs and costs associated with failure of any part of the system (in particular, transponders in a communications satellite)
- *Delay risks coverage*: To cover loss of profit or revenue, ability to service loans, expense of alternative satellite capacity, or launch services and project termination

Insurance contracts can also be arranged for third-party liabilities, which cover "innocent parties" against launch accidents or spacecraft reentry (with reference to the stipulations of the UN Outer Space Treaty of 1967 and Liability Convention of 1972); in-orbit collisions may also be covered, but fault must be proven (which is difficult in practice). Finally, coverage may also be arranged for political risks and Force Majeure (such as revocation of export licenses, confiscation of a satellite prior to launch, termination of a launch contract following civil disturbance, etc.).

An interesting quirk in the satellite business, which has important ramifications for the space insurance industry, is the lack of "product liability exposure" for satellite manufacturers beyond the point of launch. In other words, any manufacturing warranty ceases at launch and the manufacturer accepts no responsibility for failure of the manufactured hardware after that point. This represents a significant difference between satellite contracts and those for other products of technology, and refers back to the early days of high launch failure rates. Effectively, this axiom of space insurance shifts responsibility to the buyer who is obliged to arrange insurance to cover the risk of launch or in-orbit failure.

6.4.5 Process

As mentioned above, for the client, the process of arranging a space insurance policy has many similarities with any other insurance proposition; a key difference, however, is the limited number of spacecraft requiring insurance. This tends to make the policies more bespoke than, say, policies for insuring a fleet of Airbus passenger aircraft. Also, because in most cases clients have a single, expensive item to insure (e.g., a $300 m communications satellite), they tend to develop a closer relationship with their insurance broker, the manufacturer, and with the technology itself.

One of the first steps, therefore, is to engineer a mutual understanding among the key partners in the relationship: the broker needs to understand the client's requirements and how they relate to the technology; and the client needs to understand the insurance community's requirements and, among other things, how coverage relates to cost.

A key part of the insurance broker's role in the process is to ensure this mutual understanding. It typically begins with an engineering analysis of the client's satellite system, which highlights what the system is designed to do and how it does it. A key word here is "highlight"; the broking team does not need to know every minute detail of channel guard band frequency, heat-pipe routing, and solar cell adhesive, but an overview of relevant satellite platform and payload elements helps brokers to appreciate the technical risk profile and thus the client's insurance needs. For example, a client's business case may rely on providing communications services to two separate geographical service areas simultaneously, but the relative proportion of bandwidth allocated to the areas may need to be flexible. The client needs the satellite hardware designed to provide this flexibility to remain functional, but also needs to insure against the possibility that it will fail.

In the early days of satellite communications, payload design was relatively simple in that a given amplifier chain routed a single "carrier signal" (e.g., a TV channel) to a single antenna, which illuminated a given fixed geographical area. If a key part of the amplifier chain failed, a backup unit could be substituted, but there was no other flexibility and the "intended use" of the system was clear. Today, the subtleties of transponder redundancy, beam connectivity, operating mode options, and the general topic of payload reconfigurability can become so complex that, in extreme cases, it may not be feasible to test each and every combination of signal pathway through the payload.

This is where policy design becomes important. Combining insurance and engineering expertise, the broker's team designs a policy that fully reflects the engineering design of the satellite and its intended use. For a communications satellite, this typically involves the creation of so-called "loss formulae" encompassing the client's perception of the value of individual transponders or whole repeaters (collections of transponders) within the communications payload, as well as any specific technical requirements regarding antenna beam coverage and received power levels at the ground. It may also be necessary to account for a client's wish to assign different values to different parts of the payload in a "weighting scheme."

The next challenge is to present the satellite, system, and technology to the insurance underwriters in a way that will address the obligation to provide sufficient information to allow them to assess the risk, while providing encouragement to do so. Insurance underwriting is, collectively, a business that needs to make a profit; this means that, although it would clearly be counterproductive to reject all proposals, an assessment of risk must be made for each and every proposal. This begins with the technical presentation and continues with a question and answer period, which either confirms a particular underwriter's interest in covering the risk or not.

As implied earlier, space insurance differs from aerospace insurance because of the small statistical sample presented by the limited number of projects, so while aircraft insurers might rely on statistics to analyze risks satellite insurers cannot. The much larger product populations in aerospace also mean that aircraft and their engines have become far more standardized than satellites, so aircraft insurers do not require such in-depth presentations. In effect, they rely on a fundamental principle of insurance known as "utmost good faith," which means that the insured has a duty to disclose any "material information" that may affect the insurance policy. Within the constraints of commercial confidentiality, export control regimes and the like, satellite manufacturers provide much more detail for each satellite program, but also rely on insurers to do their own "due diligence" by making greater efforts to understand the risks involved. For these reasons, the leading space underwriters employ engineers or retain consultants to advise them.

Once the presentation phase is complete, the broker negotiates with underwriters in a "broking" process, which results in the placement of risk with a number of lead underwriters and other insurers that form what is known as a "following market." In this way, the broker builds financial support up to the sum required to cover the risk (say \$300 m for a satellite launch and one year in orbit), typically piecing together larger sums from the leaders and smaller ones from the followers. Contracts can then be signed and the client's satellite may be considered "insured."

One of the most important factors to bear in mind when arranging insurance is market capacity; in other words, the amount of money underwriters make available to cover space risks (which varies according to market conditions). An issue arises, for example, when two \$300 m satellites are to be launched by the same vehicle, which places a notional \$600 m risk in the market; if the market capacity at the time is only \$500 m, it will not be possible to provide full coverage.

In fact, as of late 2016, the market capacity for launch from more than 40 insurance companies worldwide was about \$750 m (Kunstadter 2016); this is somewhat theoretical, however, as it implies all underwriters will utilize their maximum capacity. Although it is becoming increasingly difficult to apportion risks geographically (because of cross border support among underwriters), the suggested split in 2014 was as follows: United Kingdom $240 m; France $157 m; rest of Europe $150 m; the United States $110 m; Asia $93 m (Kunstadter 2014). What is clear, in this regard, is the relatively high proportion available from UK-based markets (especially in comparison with the United States), which gives an indication of the importance of financial services in the United Kingdom and the position of Lloyd's of London, in particular.

It is also clear that overcapacity in the market is good for insurance buyers, because underwriters will be forced to compete for business, and it is in this aspect that brokers have the potential to save money for their clients by negotiating lower insurance rates (the rate being the percentage of the insured value charged as a premium).

A link between capacity and rates means that the two quantities tend to have an inverse relationship with a period of several years: in 2003, during the last trough in capacity (<$400 m), rates were as high as 21% for launch and 3% or so for in-orbit cover; since then, however, as capacity has risen brokers have been able to negotiate lower rates. In recent years, average rates have been historically low because of overcapacity in the market, and for 2016 they were typically between 4% and 6.5% for launch-plus-one-year and 0.4%–0.6% for in-orbit life (Kunstadter 2016); this means that it would cost an average of $16 m to insure the launch of a $300 m satellite and its first year in orbit, and perhaps $1.5 m per annum for renewal. In an average year, insurers expect to earn 80% of their income from launch policies and 20% from in-orbit coverage.

When it comes to renewal of coverage, a technical review (including analysis of satellite health reports and anomaly notifications) is conducted to assist the broking team in their discussions with insurers. If the system is generally healthy this is a relatively simple process, but if problems have occurred an analysis of anomaly resolutions—including any redesign of software and procedures—can add complexity.

If it becomes necessary to claim on the insurance policy, either for the total loss of a satellite or partial loss of its functionality, a claims analysis procedure is implemented by the broker. Although it is ultimately up to insurers whether or not a claim is paid, in full or in part, an element of the broker's role is to ease the process for its clients. Again, this is often a matter of providing the right technical information in sufficient detail to convince insurers that a claim should be paid. In this regard, underwriters will consider the "proof of loss" documentation and assess the validity of the claim, much as they would for any other insurance policy. Of course, another fundamental difference with space insurance is that the insured product cannot be seen or inspected to prove the loss; and this is another aspect of the "utmost good faith" principle of insurance.

Is space insurance a profitable venture? The answer depends on who you ask and when. Figures for 2016 show total premium income of some $630 m against claims totaling $122 m, which implies that 2016 was a very profitable year. However, when one considers that the largest ever space insurance claim totaled $406 m, it is clear that the market cannot sustain two such claims per year without repercussions (the first of which would probably be an increase in rates). In fact, the differential between premiums and claims for 2015 was much less: $725 m against $665 m, respectively.

Indeed, the space insurance market is continually on a knife-edge as far as profitability is concerned, mainly because of the limited number of insured launches and the inability to spread risk. In a market where one launch can generate 10% of annual income, but one loss can obliterate the entire annual premium income, the operative word is volatility.

6.4.6 Calculation

In cases where a total loss is not appropriate, it is necessary to calculate what is known as the "loss quantum" and this is where the partial loss formula in the insurance policy becomes relevant. If it has been properly designed, it allows figures representing the reduction in capability to be readily "plugged into" a formula to indicate the percentage of the insured amount which should be paid to the client. Thus, in the simplest scenario, if the satellite's capability can be shown to have decreased by 25%, then the claim would be 25% of the sum insured. Of course, in reality, a policy may include "deductibles," such as stipulating that a given number of transponders must fail before a claim is due. In engineering terms, this could be considered as "margin;" by analogy with household policies it represents an "excess."

Without delving into the full details of satellite partial loss formulae, it is worth repeating that the complexity of satellite insurance has increased with time. Historically, insurance claims for communications satellites were based on the loss of transponder-years (simply the number of failed transponders multiplied by the remaining part of the satellite's design-lifetime) or a loss of power or propellant, or a combination of the three.

The majority of failures were based on hardware breakdowns, solar array damage or degradation, and propellant leaks or excess usage (e.g., because a thruster valve was stuck "open"). Certain types of equipment (typically mechanical, rotating, deployable, or non solid-state parts) were known to be less reliable than certain other components and attracted the most concern among the insurance community; other, more reliable components were, to a large extent, invisible.

Today, satellite payloads in particular are far more complex and loss formulae are based on performance with reference to selected technical parameters (e.g., equivalent isotropic radiated power (EIRP), receive system figure of merit (G/T), and transmit and receive isolation). Many communications satellites incorporate separate repeaters operating in different frequency bands and serving different applications, and each will make a given contribution to the "intended use" of the satellite; and, of course, this contribution has a financial element. The partial loss formula attempts to reflect the increased complexity of the satellite payload, while incorporating the arguably more mundane elements associated with platform failure (such as power and propellant loss).

Although the platform elements for imaging satellites are similar (notwithstanding the different environment of the lower altitude orbit and consequent operational differences), the payload element is entirely different; it is essentially an imaging system (camera) as opposed to a radio-relay system (receiver/transmitter). In this case, a typical loss formula is based on a number of parameters grouped under two headings: collection capacity and image quality (effectively addressing both quantity and quality performance aspects).

Collection capacity is defined as the ability to collect, store, and downlink imagery and includes parameters measuring data collection rate, on-board memory capacity, and the capability to downlink data to a receiving earth station. It may also include a measure of the satellite's agility (how quickly it can point to another target) and of detector performance (the number of operational elements in the imaging device). Image quality is typically determined with respect to four performance metrics: accuracy of pointing and geolocation; ground sample distance (an image resolution measure); signal-to-noise ratio; and modulation transfer function (a measure of image sharpness).

Using such loss formulae, and derivations of them, it should be possible to design a policy for any spacecraft, whether it is a single commercial communications or imaging satellite, a constellation of such satellites, or indeed a commercial manned space capsule (such as those being designed to deliver crews to the International Space Station).

Even a cursory analysis of the development of the space insurance market since its beginnings in 1965 shows the degree to which the market has evolved to address developments in technology and alternative risk scenarios. Today, more than ever, a clear understanding of space technology and an ability to assess and prioritize potential risks is important for the design of insurance policies, the analysis of spacecraft health reports for policy renewal, and the assessment of any claims. Moreover, considering the plethora of possible future developments in the space industry and their various disparate needs, the space insurance business is not about to get any simpler.

6.5 FUTURE DEVELOPMENTS

Although it is difficult to predict the future for space insurance, it seems likely that there will be a continuing, and arguably growing need for this type of business risk mitigation.

That said, some of the larger satellite operators, with dozens of satellites in their fleets, are able to "self-insure" to an extent: if one of the fleet fails they simply replace it with another, perhaps underutilized, satellite until a replacement can be launched. Indeed, the business model for most LEO and MEO constellations involves having a number of in-orbit spares. Figures for 2016 show that "23% of GEO operators buy little or no in-orbit insurance, representing 48% of the commercial in-orbit fleet" (Kunstadter 2016), which makes this a significant challenge for insurers seeking to spread risk across a large population.

It remains the case, however, that smaller operators are unlikely to be able to withstand the loss of a key (or even single) asset without severe financial repercussions and will always require insurance. Even for the larger operators, the launch phase continues to be an inherently risky part of the process and will need to be insured.

The commercialization of space beyond communications and imaging satellites has progressed somewhat more slowly than expected, partly because of the technical challenges but also because of the financial risks involved. For this latter reason, in particular, when commercial space stations, asteroid mining, and space tourism finally do appear they will have to be supported by insurance.

Moreover, any significant increase in the use of the space environment, whether in low Earth orbit or beyond, brings with it the potential for a degradation of that environment. In LEO, the risk of a self-sustaining "cascade effect" as a result of multiple debris collisions was hypothesized as long ago as 1978, and since then, an increase in the orbital debris population has been well documented (Kessler and Cour-Palais 1978).

Although, to date, there have been no space insurance claims as a result of proven debris collision damage, there *has* been speculation that debris was the "most likely cause" of some on-board failures; the problem is proving culpability in relation to such remote assets. That said, the number of reports of avoidance maneuvers conducted by low Earth orbiting satellites is on the increase and the question of the long-term sustainability of some highly populated orbits is in question. Moreover, the sustainability of insurance as a risk mitigation measure is, ultimately, also in question. The trend toward 10 cm "cubesats" (weighing anything from 1 kg upward) is seen by some as a potential source of orbital debris, especially since most of the smaller satellites have no propulsion system for orbital control and deorbiting. In a worst-case scenario, a single collision between a cubesat and a larger satellite could produce a cloud of debris, which would impact yet other spacecraft bringing about a chain reaction effect that could make an entire orbit unusable.

Although the majority of commercially insured satellites reside not in LEO but in the higher geostationary orbit, where debris populations are lower, the ultimate risk of a debris problem here is potentially even greater because there is only one geostationary orbit. Current best practice involves boosting geostationary satellites to a "graveyard orbit" 250–300 km above GEO, but it is not universally followed as a debris mitigation measure. Moreover, moving satellites to the graveyard should be viewed only as a temporary measure because those satellites are, by definition, uncontrolled (as they have no remaining propellant) and will continue to pose a collision risk for as long as they remain there. Figures suggest that in 50 years or so there could be up to a thousand uncontrolled satellites in the graveyard orbit (Williamson 2002).

The increasing risk presented by orbital debris—not only in the various Earth orbits but also in orbit about the Moon and Mars—is but one aspect of the thesis characterizing space as a "fragile frontier" (Williamson 2006). Apart from any moral or ethical issues related to environmental damage, the relative lack of any self-repair capability in orbital environments and the total lack of the same in the lunar surface environment mean that commercial development can lead to permanent degradation or damage.

The ultimate risk in this regard is that of the long-term sustainability of the space environment and thus of space-based assets, commercial or otherwise. The challenge in the second half-century of the Space Age includes the protection of certain, yet-to-be agreed upon, aspects of the space environment for scientific study and the preservation of other natural assets for use by future generations. This remains a work in progress.

ACKNOWLEDGMENT

The author would like to acknowledge the kind assistance of David Wade of the Atrium Consortium and Russell Sawyer of Willis Towers Watson in ensuring the accuracy of this contribution.

REFERENCES

Gubby R., Wade D., and Hoffer D., 2014. Preparing for the Worst: The Space Insurance Market's Realistic Disaster Scenarios. IAC-14, D6.2-D2.9, 7×22494 (IAC Toronto).

Hill S. M., 1983. *Space Communications & Broadcasting*. 1, 393–403. Amsterdam: Elsevier Science Publishers B.V. (North-Holland).

Kessler D.J. and Cour-Palais B.G., 1978. Collision frequency of artificial satellites: The creation of a debris belt. *J. Geophys. Res.* 83(A6), 2637–2646.

Kunstadter C.T.W., 2014. Market Update, World Space Risk Forum (Dubai), May 2014.

Kunstadter C.T.W., 2016. Space Insurance Update, World Space Risk Forum (Dubai), November 2016.

Lloyd's, 2017. *Realistic Disaster Scenarios: Scenario Specification*, January 2017, EM 196 V1.0. Available at: https://www.lloyds.com/the-market/tools-and-resources/research/exposure-management/realistic-disaster-scenarios/scenario-specification-2017 (Accessed: April 5, 2017).

Todd D., 2016. Early Flight Reliability, World Space Risk Forum (Dubai), November 2016.

Wade D., 2013. Space weather: A space insurance perspective. *Science in Parliament*, 70(2), (Whitsun 2013), 22–23.

Williamson M., 1990. *The Communications Satellite*. Bristol & New York: Adam Hilger/Institute of Physics (ISBN 0-85274-192-8).

Williamson M., 2001. *The Cambridge Dictionary of Space Technology*. Cambridge: CUP (ISBN 0-521-66077-7).

Williamson M., 2002. Protection of the Space Environment: The First Small Steps. COSPAR02-A-01364/PPP1-0014-02 (World Space Congress, Houston).

Williamson M., 2006. *Space: The Fragile Frontier*. Reston: AIAA.

Williamson M., 2016. Commerce at the Cape, Engineering & Technology, August/September 2016, pp. 46–49.

CHAPTER 7

Space Sustainability

Christopher D. Johnson

CONTENTS

Space technology is a foundational component of our globalized and interconnected world. Space-based capabilities such as Earth observation, position navigation and timing (PNT), telecommunications, and a host of other applications enable the daily activities of individuals, corporations, and states. The other chapters in this volume examine possible hazards in and from the space environment, including the threat of asteroids and comets impacting the Earth; space weather and space debris threatening activities in space; the extraction and use of space resources; ground-based conflict involving space assets; and the threat of conflict in outer space itself. Additionally, international politics and economic trends, technology development, and other issues have always and will continue to influence space activities.

7.1 IRRESPONSIBLE STEWARDSHIP

Humankind has reaped great benefits and profits from the resources we discover, claim, and use. However, these shared assets have a tendency to be mismanaged, polluted, and ultimately exhausted. Whether inside state territory or in shared common areas out of states, society does not have a good historical track record in responsible stewardship of the environment. All too often, short-term goals and gains overrule long-term considerations and values. Once the trees on one tract of land are completely logged, we move on to fresh sources of lumber. Once one type of fish species is caught to the point of population collapse, we trawl deeper for sea life previously considered too unattractive or unmarketable.

A look at the vast common areas outside of state territory such as airspace and the high seas shows mixed results. It's true that aviation has connected and transformed the modern globalized world, but aviation is not without its long-term impacts on the environment, including its contribution to atmospheric pollution and climate change. On the high seas, the maritime industry underpins the industrialized world. However, commercial traffic and activities on the high seas are often grossly underregulated and mismanaged (Langewiesche 2004). The health of the high seas and the life within it are bearing the brunt of our ever-growing ambitions and suffer direct harm from our neglect and avarice. In these shared spaces, the existing regimes persist because (at least for now) benefits continue to accrue to most stakeholders, and most risks and long-term consequences seem remote or will be borne by others in the future. We can and should learn from and improve upon the governance regimes of the shared domains of international airspace and the high seas. It should be our ambition that the regime(s) for outer space be more coherent, more inclusive of both short- and long-term views and values, better adhered to, and more effective in fostering the future we want.

7.2 CONCEPTUAL HURDLES

For those who work in the space industry or who are otherwise connected to outer space activities, space is the topic upon which our thinking and energy is continually focused. But despite this, outer space is a region to which almost none of us will ever travel to or gain firsthand experience with. The space environment has many unique qualities and characteristics which make it especially difficult to conceptualize in a comprehensive and coherent fashion, and space is a region whose immensity is difficult to fathom. While spacecraft travel fast compared to objects in our everyday experience, the distances between useful orbits and interesting destinations is likewise vast. Because of the strangeness of the space environment compared to what we can easily imagine, concepts and issues such as space weather, space debris, and conflict in outer space are as outside our everyday frame of reference as imagining the size and qualities of molecules or their atomic structures. Those discussing space often rely upon analogies to other domains, including aviation, the seas (both above and below the water line) and even cyber space. But all analogies break down after a certain point. And because of the unfamiliar nature of the space domain, biases and misperceptions can accumulate to distort our perceptions and assessments.

For example, in thinking about space debris and the severity of the space debris problem, two impressions may arise. The first is acknowledging that (at least in some orbits) the space debris problem has reached the "tipping point," and the Kessler syndrome has begun. Debris-on-debris collisions will continue to increase the number and amount of space debris, even without additional launches of new spacecraft (Krag 2017) (Figure 7.1).

Therefore, space debris is an issue which must be faced, and every planned launch and subsequent operation in space must perform the necessary tasks to ensure that space debris remediation activities are conducted. These tasks include performing prelaunch conjunction analysis with the existing catalogs of space debris, and provisioning enough fuel for

FIGURE 7.1 Space debris effects in the geostationary orbit. The geostationary ring is one of the most congested regions in space, and the average distance between two objects here is only 190 km. (Adapted from ESA/ID&Sense/ONiRiXEL, CC BY-SA 3.0 IGO, http://www.esa.int/spaceinimages/Images/2017/03/Space_debris_effects_on_the_geostationary_orbit.)

on-orbit collision avoidance and end-of-life maneuvers. Additionally, new projects must still accept a degree of risk for those pieces of space debris too small to be tracked or even modeled with available technologies. In order to receive launch authorization, operators might also be required to guarantee to not create any additional debris with their proposed activity, and to comply with space debris mitigation regulations.

The second and opposing impression may form based less on technical data and more on first impressions. When you consider the vastness of the space environment, even in the various Earth orbits, it does not seem that these places are filled with debris. One might think that because there is literally nothing larger than outer space, how can it be that space is actually "contested, congested, and competitive" (as is sometimes asserted)? (Figure 7.2).

Nevertheless, the data and facts as explained by the technical experts seem certain, and serve to remind us of the reality of the various issues and challenges which actually exist. Chapter 5 in this volume on "Space Debris: Risk and Mitigation" should serve to disabuse those who hold the notion that they need not worry about space debris. No matter if these facts are inconvenient, bothersome, or space sustainability is costly and time-consuming, any future space project will confront these hazards.

FIGURE 7.2 Low earth orbit. Images like this, showing the Earth from low earth orbit (LEO), show space appearing pristine and unpolluted. (Adapted from ESA/NASA, https://www.nasa.gov/sites/default/files/thumbnails/image/23756761750_088b6f670d_o.jpg.)

The challenges discussed elsewhere in this book may be helpfully subdivided into two main categories: naturally occurring hazards such as asteroids, comets, space weather, and naturally caused radiofrequency interference; and problems created by humankind's use of outer space: space debris, human-caused radiofrequency interference, and so on. Some of these human-caused issues lend themselves to analysis using economics concepts such as common pool resources, free rider problems, and the tragedy of the commons. Another troublesome aspect of humanity's use of outer space is the possibility of space being a domain of conflict. We might therefore also consider these thorny issues using traditional international relations and collective security approaches.

7.3 SPACE SUSTAINABILITY DEFINED

However, the clearest and most focused prism to view all of these challenges is through the lens of space sustainability, which considers the interests and needs of current, planned, and potential activities and uses of outer space, and reflects on whether the present state of affairs and the trends of existing activities either meet or fail to meet these needs.

Some may contend that sustainability is an undefined term, or perhaps so loosely defined so as to prevent coherent analysis and action. However, a closely related concept to "sustainability" is "sustainable development."

Sustainable development is "development that meets the needs of the present without compromising the ability of future generations to meet their own needs" (Brundtland Commission 1987). This is a nuanced and perhaps difficult formulation, because it requires us to imagine interests which are remote to us in time, and held by others in the future. This difficulty has made contemplating and incorporating long-term interests a challenge in a variety of fields.

Particular to the space domain, space sustainability is ensuring that all humanity can continue to use outer space for peaceful purposes and socioeconomic benefit both now and in the long term (Secure World Foundation 2014a). Space sustainability is used to refer to "a set of concerns arising out of the realization that near-Earth orbit and the electromagnetic spectrum are limited natural resources that are under increasing pressure from the steady growth in the number and diversity of space actors" (Martinez 2016). As stated, space sustainability looks at existing conditions (both natural and human-caused) and considers the future they implicate. It asks, "With the ways things are going, what will the future look like?" If we do not harden our satellites to protect them from space weather, making our systems resilient and/or redundant, what vulnerabilities will our space systems—and the ground systems linked to them—be exposed to? If we keep launching spacecraft with boosters, rocket stages, and fairings to be left in orbit, what will the space environment look like in the future?

Space sustainability also asks us to look at ourselves and our actions while asking, "What if everyone started doing what I'm doing? Would I want that?" And "If everyone followed my example, how would I myself be affected?" It asks us to consider the repercussions of asserting explanations and excuses for diverging from the rules. One might protest "The rules don't apply to me, because I'm in a special situation" or "I can't adhere to the rules this time, it's too costly, burdensome, or unfair," or even "In the past, other actors didn't have to adhere to the rules, and they were free to pollute the space environment. I should get to act like they did, before the rules were broadly imposed on everyone." The legitimacy of these excuses can always be tested by asking what the impact would be if others were also given the freedom to assert excuses when they wished to.

7.4 THE CURRENT ENVIRONMENT

In developing true space sustainability, the first task is an understanding of the current space environment, including uses and abuses. This book provides ample substance for such an analysis. Other chapters focused on natural risks such as asteroid and cometary impact hazards, and space weather. These phenomena will persist, but the risks they present can be guarded against. The same types of rationales which lead one to get health and life insurance or other protections from low probability and high severity risks also applies to these naturally occurring space hazards. Space sustainability encapsulates recognizing and planning against these natural space hazards.

Other chapters covered space debris, legal complications from commercial space and celestial resource use, and the ever-present and evolving threat of bad actors and conflict in space all stem from humankind's use of outer space. These issues are perhaps harder to address, while also not likely to disappear completely. They will require both technical advancements and political, legal, and regulatory developments at international, national, and even contractual levels. Having considered these other chapters on the various hazards to space activities, one can arrive at their own formulation of the present state of space activities, and some of the most urgent and troubling concerns confronting current actors in space activities. Space operations are already affected by these issues, and space sustainability is a topic gaining wider attention.

7.5 EXISTING EFFORTS

For the past few decades, a growing awareness of our responsibility for caretaking of the natural environment has dawned on society. That realization has been unequal across constituencies and demographics, to be sure. It could have been pursued faster. And, if we are to avert further degradation of the space environment, our actions for space sustainability must continue apace—if not accelerate. However, the concept of sustainability and the content of what that phrase means has slowly but steadily been incorporated into many fields and pursuits, including space.

7.5.1 Space Debris Mitigation Guidelines

Space debris has been a known problem for quite some time. Formally established in 1993, the Inter-Agency Space Debris Coordination Committee (IADC) is a body comprised of national space agencies, including the American, British, Canadian, Chinese, French, German, Indian, Italian, Japanese, Russian, South Korean, and Ukrainian space agencies, along with the European Space Agency (ESA). The IADC began work and analysis on space debris on a technical level and formulated the first version of space debris mitigation guidelines in 2002, revising them in

2007. The IADC Debris Mitigation Guidelines call for current and planned space missions to adhere to certain rules, including a 25-year deorbit rule, passivation of inactive satellites, provisioning for on-orbit maneuvers, and performing collision avoidance (Inter-Agency Space Debris Coordination Committee 2007). See Chapter 5 for further technical details of these mitigation guidelines.

These IADC Debris Mitigation Guidelines, which followed a "bottom-up" approach from the technical and operational level, were complimented with a more political and state-to-state set of guidelines developed within the United Nations Committee on the Peaceful Uses of Outer Space (COPUOS) in 2007. The COPUOS space debris mitigation guidelines consist of a mere seven guidelines and are rather short in nature. As an example of their brevity, Guideline four reads: "[r]ecognizing that an increased risk of collision could pose a threat to space operations, the intentional destruction of any on-orbit spacecraft and launch vehicle orbital stages or other harmful activities that generate long-lived debris should be avoided" (United Nations Office for Outer Space Affairs 2010).

However, both the IADC and the COPUOS guidelines are nonbinding sources of guidance, and neither strictly impose compliance, nor form mechanisms for punishment of noncompliance. The COPUOS guidelines explicitly state that they "are not legally binding under international law. It is also recognized that exceptions to the implementation of individual guidelines or elements thereof may be justified, for example, by the provisions of the United Nations treaties and principles on outer space" (United Nations Office for Outer Space Affairs 2010).

Regardless of their status under international law, the impact through compliance and observation matters more for space sustainability. Significant adherence through implementation on a national level, both with national space legislation and observance in national programs and policies, has been undertaken by a number of states. However, the success rate of spacecraft compliance at the end of their operational lifespan with guidelines, called Post Mission Disposal (PMD), is tracked by the European Space Agency. Holger Krag, head of ESA's Space Debris Office, has written that since 1990, "The annual PMD success rate has wavered around a level of 60%, which means that roughly 46 satellites get "stranded" in space each year" (Krag 2017). It appears that efforts toward debris mitigation and compliance with guidelines are a middling success, which must be improved upon. So far, avenues to increase compliance remain mostly unpursued.

7.5.2 UN Group of Governmental Experts

From time to time and as a response to geopolitical trends and issues, the United Nations will establish a Group of Governmental Experts to consider emerging issues which might impact international peace and security. In 2010, UNGA Resolution 65/68 on Transparency and Confidence Building Measures in Outer Space Activities requested the UN Secretary-General to establish a Group of Governmental Experts (GGE) specific to outer space transparency and confidence building measures, and to report back to them by the end of 2013. Experts from fifteen countries were organized in the group in 2012 to study and reflect on the utility of using transparency and confidence building measures (TCBMs) in the space domain as a means to alleviate existing and possible tensions. TCBMs are a useful tool in international discourse and diplomacy, especially in areas with security concerns, and can be used as foundations for future multilateral initiatives (Hitchens 2016). And, as TCBMs have the potential of reducing the risk of misapprehensions and escalation between actors, they have a significant and crucial role to play in fostering space sustainability (Secure World Foundation 2014b). Unlike the IADC and efforts tied to COPUOS, the GGE also considered security aspects of space activities.

The GGE on TCBMs met three times during 2012 and 2013, and its final report was transmitted by the UN Secretary-General to the UN General Assembly First Committee in October 2013. The final report of the GGE on TCBMs in space is short in nature but quite broad in scope and potential influence. The GGE report called for information exchange (i.e., transparency) on national space policy and goals, military space expenditures, and activities in outer space (including orbital parameters, possible conjunctions, natural space hazards, and planned launches) (United Nations 2013). The report also called for notifications on risk reductions such as scheduled maneuvers, uncontrolled high-risk reentries, emergency situations, and intentional orbital breakups. Last, the GGE report encouraged voluntary visits to launch sites and command and control centers, and demonstrations of space and rocket technologies. Since the finalization of the report, implementing the GGE has been a topic for discussion but actual concrete steps remain to be taken by individual states and other actors (Hitchens 2016).

7.5.3 Long-Term Sustainability Guidelines

In 2009, COPUOS established the Working Group on the Long Term Sustainability of Space Activities (LTSSA) within the COPUOS Scientific and Technical subcommittee. In 2011, the LTSSA Working Group adopted terms of reference for its work, and established four Expert Groups to consider various aspects of space sustainability, including discussing space debris, space situational awareness, space weather, regulatory regimes and guidance for new actors in space, and space utilization supporting sustainable development on Earth. State Members of COPUOS nominated experts to work in the expert groups, and work began on a set of guidelines for the long-term sustainability of space operations.

As of 2018, the long-term sustainability guidelines are still in development. A number of guidelines have already been finalized, and others are still open for debate and refinement within COPUOS. However, like the previous guidelines on space debris, the long-term sustainability guidelines are of a nonbinding nature. They merely reflect existing best practices as already employed by states, and on topics and activities already undertaken in space activities. They therefore do not further the progressive development of norms into new activities or areas. They do have merit as a consolidation of best practices promoting space sustainability, which can be used as guidance for states. Once completed, the long-term sustainability guidelines are expected to be included as an annex in the annual United Nations General Assembly resolution on space (Martinez 2016).

7.5.4 Small Satellites, Megaconstellations

Current issues impacting space sustainability include the space debris problem and governance efforts to mitigate it—including intentional and unintentional creation of space debris, and Post Mission Disposal adherence rates. However, innovation continues, as does planning for new project and programs. In recent years, the small satellite revolution has opened the playing field to new actors in space. It will continue to bring many more actors, novel concepts, and pioneering technologies to the space domain. A related trend to the changes that small satellites will bring to the space environment is the numerous and varied megaconstellations being announced and designed. Constellations of hundreds and even thousands of small satellites operating in a constellation under the control of a single private entity will stress and challenge the space environment and the international and national governance regimes surrounding it (Figure 7.3).

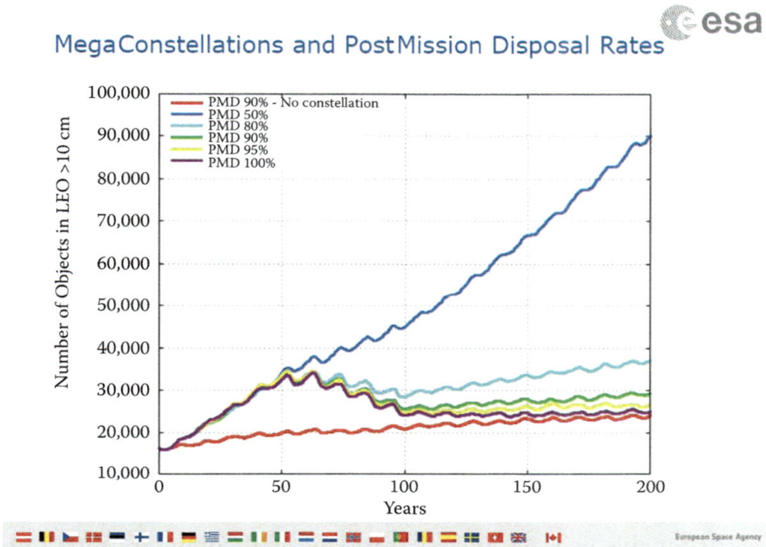

FIGURE 7.3 Megaconstellations, Post Mission Disposal (PMD), and the possible future growth of space debris. Simulation results for the evolution of the population of objects greater then 10 cm, with background space traffic implementing mitigation measures at a 90% success rate, and with the effects of a sample megaconstellation implementing mitigation measures at various success rates. (Adapted from Holger Krag, ESA Space Debris office, http://www.unoosa.org/documents/pdf/copuos/stsc/2016/tech-09E.pdf.)

7.6 THE LIKELY FUTURE

Based on the present field of space activities and past trends (e.g., the historical growth of space debris in Figure 5.2), we can extrapolate into the future. Any realistic chance of ensuring space sustainability requires we do so in an honest, realistic, and frank fashion.

Broadly looking at all global space activities, we are already seeing both an increase in the number of actors and the variety of activities in space. We should anticipate more actors and an increasing diversity of these actors.

We can anticipate that future actors in space will be doing everything they are currently doing now, and begin to do many of the things they are now realistically planning to do (on-orbit servicing, megaconstellations, etc.). Also, we can anticipate that there will be space activities and technologies which we have not yet imagined. In other words, we can anticipate that we will be surprised by innovation and ingenuity from the space sector, in addition to new entrants and activities being announced around the world. This rate of change, innovation, and disruption will continue to accelerate.

7.6.1 Attitudes Toward Governance

In addition to all these potential space activities, we should also consider attitudes toward policy, law, regulation, governance, and more general attitudes toward space sustainability as a necessary concept worth pursuing. Do actors in the space domain already know about space sustainability issues? In the future, will actors be willing to alter their behavior to aid in space sustainability, or will they be looking to get by and rely on other people altering their behavior (i.e., the free rider problem)? Will there be a better framework for governance in the future?

Future space activities will involve a growing diversity of actors and novelty of activities, and an increase in public-private partnerships and international joint ventures. Will these changes create conditions for more or less sustainable uses of outer space? And, will compliance with norms which foster sustainability be increased in the future? This last option is more difficult to predict, but it has the effect of adding in some modesty to our prognostications to the future.

7.6.2 Governmental Oversight

A state's capacity to govern space activities should also be considered. Under Article VI of the Outer Space Treaty (OST), states are both internationally responsible and potentially internationally liable for their national activities in outer space, including the activities of nongovernmental (private) actors. In a manner which is unique in international law, damage caused by private actors in space is the direct liability of the launching state(s) of those private space activities. Additionally, Article VI requires that the activities of "nongovernmental entities in outer space… shall require authorization and continuing supervision by the appropriate State party to the Treaty."

Fulfilling oversight duties under Article VI requires administrative competence, effort, and the outlay of resources. In a future of expanded and widely diverse space activities conducted by a myriad array of nongovernmental actors, what will be the organizational and bureaucratic burden to effectively license, authorize, and supervise all of the private actors? What does "authorize and supervise" mean in terms of practical implementation? What minimum level of oversight is required by the OST? And, to ensure space sustainability, is a minimum level of oversight sufficient? As the ratio of private actors increases in relation to purely national governmental space activities, can states continue to effectively oversee and regulate all these private activities? Rather than individual payloads and launches, will governmental agencies license entire constellations and whole classes of space activities?

7.6.3 Free Riders and Bad Actors

Last, in contemplating the likely future of activities in outer space, we should also recognize that there will always be free riders and bad actors in any system. There will always be those who break the law. This does not make the law cease to exist. Bad actors and free riders do not obviate the need for governance systems. Rather, they bolster the need for governance systems, because clear rules—and the mechanisms for their enforcement—have a deterring effect on most actors who might be tempted to defect from or disobey the rules, or otherwise place their own needs beyond the impact their defections from the rules would have on others. Additionally, even when actors violate the rules, they often protest that their actions do not actually violate them, or alternatively, that extenuating circumstances exist which excuse them from adhering to them. In so doing, despite their violation in the eyes of others, bad actors simultaneously reinforce both the existence of the rules and the importance of adhering to them. This is obviously better than a situation with a complete absence of rules.

There is also a paradox of altruism. In a system where most adhere to the rules, everyone benefits. However, occasional bad actors and free riders can stand to benefit more in these environments—environments where everyone else abides by the rules. This may be true, but it is still preferable to a system where rules are absent.

Additionally, as more and more actors in the space field continue to have a stake in favorable outcomes under effective and sustainable governance regimes, and would have something to lose from the degradation of a working system, there will be more adherence to and respect for the rules. This is especially true in the space domain where space debris threatens indiscriminately anyone using useful orbits.

7.7 A WORSE FUTURE

Having spent time thinking about some of these characteristics of the possible future in space according to current trends, in order to sharpen our thinking, we should envision both good and bad futures. In other words, we should imagine worst possible future scenarios and outcomes, and best possible future scenarios and outcomes. Luckily, rather than shaky predictions and future forecasting, imagining good and bad futures is not principally dependent on forecasting data as much as contemplating futures which either meet or fail to meet our needs for peace, security, stability, equality, technological progress, and other values. Once these are imagined, we can reflect on how likely it is that these scenarios come into existence, and more importantly what trends, governance systems, and practices would allow, assist, or prevent them from becoming reality.

7.7.1 Kessler Syndrome

A bad future in space is one where numerous problems exist and they are not managed or addressed. Imagine a future where the population of space debris is so numerous, unmanageable, and untrackable that entire orbits are so congested that it is economically unfeasible for most actors to use them. The Kessler syndrome in those orbits has debris colliding with other debris, further exacerbating the problem. Orbits cannot even be cleaned-up through remediation and capture technology. Removal of space debris might be limited by technological impediments, or because of political impediments, including the refusal of launching states to permit their nonfunctioning space garbage being captured by others. Because certain orbits are so polluted with debris, capabilities depending on these orbits are severely impaired. For example, Earth observation satellites use Sun-synchronous orbits (SSO) between 700 and 900 kilometers, but this orbit is susceptible to lasting debris. Spacecraft left at geosynchronous Earth orbit (GEO) will also likely remain there in perpetuity. In a worse future, rampant debris in these regions will leave access to these orbits severely degraded, with resulting loss of service in space applications.

7.7.2 Radiofrequency Interference

Radiofrequency interference degrades our use of space. Unintentional interference—caused either by naturally occurring space weather or through the failure to coordinate our space services through national and/or international mechanisms—means that users suffer loss of signals and can neither use the transponders on satellites for their intended applications nor communicate with their satellites to maneuver them out of harm's way. Intentional interference, including jamming, is done for a variety of purposes including national military jamming of foreign space assets and jamming of private commercial satellites. Satellites can share both private commercial and national transponders, and jamming for both PNT and satellite communications is possible.

In a worse future, both may suffer from intentional interference with troubling regularity. Enforcement to stop or prevent future jamming is ineffective and bad actors get away with degrading and denying other actors from communicating with their satellites. The use of space assets for critical functions is therefore rerouted through ground systems. Political tensions fester over disputes borne from radiofrequency interference issues. Additionally, possibilities to reduce interference are slim, and avenues for progress are not pursued.

7.7.3 Stifled Technology Development

In a worse future, emerging technologies never develop to their full potential. These might include useful applications such as satellite servicing, which is proximity and rendezvous operations of existing satellites—either to pilot them once their fuel is spent, to place them in correct orbits, to repair them, or to remove them from orbit to reenter the atmosphere or to place them in graveyard orbits (also called end-of-life services). However, because satellite servicing uses technology which could be seen as threatening to space systems, so much political apprehension accrues surrounding its development that this useful advanced technology never sees viability and deployment. Without satellite servicing technologies, satellites at the end of life are left in orbit, and new satellites must be built and launched to replace them. Satellites placed in the wrong orbit cannot be correctly re-orbited, and satellites which remain useful except for the exhaustion of their fuel supply are forfeit and become debris.

Technologies and ambitions for the use of celestial resources might never be realized. While Article II of the OST explicitly prohibits national appropriation, this provision is an artifact of an era of cold war tensions and rapid global decolonization. Intended to prevent a colonial grab for territory, Article II was never meant to prevent missions to asteroids and large planetary bodies from harnessing the resources found there for the creation of a deep space infrastructure and economy. In a worse future, Article I of the OST, enshrining freedom of access and use, is outweighed by concerns that some states and companies are seeking a monopolistic "gold rush" over resources, and therefore the use of celestial resources is considered to be clearly prohibited by Article II, which is deemed to contain no gaps or exceptions and to be comprehensive in its prohibitions. As such, no resources in space may be brought back to Earth or used for advanced missions either for producing chemicals for water, fuel, material for manufacturing, and for habitats and shielding. Seeing a regime this restrictive of what they view is their freedoms in space some states seek to either amend the OST or withdraw from it outright. As with interference, tensions over celestial resource use fester between states, driving mistrust and apprehension.

7.7.4 Planetary Defense

Article IV of the OST requires that states do not place "in orbit around the Earth any objects carrying nuclear weapons or any other kinds of weapons of mass destruction, install such weapons on celestial bodies, or station such weapons in outer space in any other manner." This prohibition, a cornerstone of the OST, was intended to prevent outer space from falling victim to a cold war era arms race. It successfully prevented the weaponization of outer space during heightened tensions between nuclear armed superpowers. However, in the context of planetary defense, where an asteroid or comet could threaten societies and states, or even life on Earth, the question has been posed that defending against an asteroid threat requires the deployment of nuclear weapons into outer space, an act formally outlawed by the OST. This ambiguity and tension between a weapon and a nuclear device for planetary defense means that asteroid threat plans become politically unsustainable, and no state is willing to risk its planetary defense actions being perceived as threatening or hostile. In a worse future, the persistent issue of planetary defense is left unaddressed.

7.7.5 Conflict in Space

Last, conflict in outer space becomes a reality. When tensions break out on Earth, a state's national space assets used to support ground operations are considered as a legitimate target and therefore implicated in the ground war. Conflict on Earth involves the space domain as merely another domain to wage war in or through. And as destroying a satellite doesn't directly implicate loss of human life, targeting a satellite is permissible under the law of armed conflict's rubrics for distinction and proportionality. In fact, during a conflict destroying foreign space assets is chosen before targeting ground infrastructure. The resulting problems of space debris are thereby worsened. However, while physical damage incurs liability under space law, under the law of armed conflict this intentional destruction of foreign space objects remains permissible and does not give rise to liability concerns.

7.7.6 Odds of a Worse Future

These possible worse future outcomes may be considered and assigned more or less likely according to the views and opinions of each reader. The point is that a worse future can be imagined, one where space sustainability is failing and where tension, mistrust, and underutilization is more common than peace, cooperation, and development. Additionally, consider a future where sustainable governance is never achieved. As issues such as space debris and novel technology development continue to impact space activities, international and national initiatives fail to assist in fostering sustainability. With no clear and workable rules, and states less willing to adhere or even develop them, no governance framework is implemented to mitigate these space hazards. Unwillingness by some states to develop any types of governance frameworks may result in a future where no states are willing to collectively address these shared problems. The ingredients for such a future are already in place.

7.8 A BETTER FUTURE

We might also imagine a better future, where many of the challenges of space sustainability are squarely faced, tough decisions are made, and progress happens because actors today and in the future are able to successfully integrate their short-term interests and objectives with their long-term interests and values. They are able to develop coherent and implementable governance systems where actors trust each other and mutually exchange obligations which restrict their selfish unilateral actions in favor of collective gains. The "Thucydides' Trap," where historical major powers cannot accommodate the interests of emerging major powers in a stable fashion is avoided in the space context. The view that space war is inevitable and the preparation for this "inevitable" war is reoriented toward avoiding inevitability and creating sustainable norms and avenues to de-escalate from tense situations. Actors in the space domain realize that if conflict were to break out in space, this precedence would be disastrous for the long-term sustainability of the space environment.

In this better future, natural hazards from space such as space weather and asteroid threats remain, but these high-impact low probability occurrences have been given adequate time, attention, and resources. Space and ground systems are capable of withstanding even the most severe space weather events, and the possibility of Near-Earth Objects (NEOs) threatening the Earth has caused lawmakers and space agencies to stand-up preparedness programs. Toward that end, efforts such as the (already existing) International Asteroid Warning Network (IAWN) and the Space Mission Planning Advisory Group (SMAPG) continue to successfully coordinate national threat detection efforts and asteroid deflection and response contingency plans.

For human-created challenges to space sustainability, a better future will see resources allocated to develop space debris remediation technologies. These technologies can both deorbit the largest pieces of historical space junk and remove the smaller bits of trash. Such technologies might include lasers, gels, foams, and nets. Additionally, the political and security issues are solved with international transparency and confidence-building measures which alleviate concerns. Potential liability issues are solved with correctly balanced regulation and contractual arrangements (such as waivers of claims) between actors.

7.9 A BROADER VIEW

The words we use shape how we think about the topic. This seems common sense, but it bears repeating and remembering as a check on the soundness of our beliefs and conclusions. Outer space is sometimes referred to as "commons," a "global commons" or an "international commons." Like the high seas, international airspace, and the Antarctic, outer space exists outside of traditional state territory. But is the space domain itself a "commons?"

7.9.1 Province of All Mankind

Article I of the OST uses the phrase "province of all mankind," stating "[t]he exploration and use of outer space, including the Moon and other celestial bodies, shall be carried out for the benefit and in the interests of all countries, irrespective of their degree of economic and scientific development, and shall be the province of all mankind." Many take this sentence to mean that space is therefore the province of all humankind. However, the subject of that sentence is "the exploration and use," rather than merely "outer space." As such, it is the *activity of exploring and using* which is the subject of this clause of the article. The freedom to explore and use outer space is the province of all humankind, rather than the place itself.

7.9.2 Common Heritage of All Mankind

Additionally, some conflate "province" with "common heritage." The phrase common heritage is found in the 1979 Moon Agreement, whose Article 11.1 states "[t]he Moon and its natural resources are the common heritage of mankind... ." Additionally, Article 1.1 of the Moon Agreement states that "[t]he provisions of this Agreement relating to the Moon shall also apply to other celestial bodies within the Solar System, other than the Earth, except insofar as specific legal norms enter into force with respect to any of these celestial bodies." Consequently, "common heritage" can be stretched to apply to the entire Solar System (besides the Earth). However, as of January 1, 2018, the Moon Agreement has a mere 18 states parties. None of them are major spacefaring states. With 193 Member States of the United Nations system it means that 175 states (90% of the total) are not a party to the Moon Agreement. As such, it is possible to see the Moon Agreement, opened for signature for close to 40 years (1979), as a failed exercise in treaty making for outer space. It certainly does not apply to the major spacefaring states, and even parties that are signatories will likely never ratify the agreement. Consequently, the phrase "common heritage of mankind" should not be used in serious discussions of future space activities.

The phrase "province of all mankind" therefore applies to space activities. The exploration and use of outer space is the province of all mankind. More equally, this exploration and use is the province of all humankind. But further analysis of that phrase is necessary. As discussed above, it is the "exploration" and "use" as undertakings, free for all to attempt, and with the prior consent of states not required. This therefore means that the physical domain of outer space, both the celestial bodies and void space itself, is not the province of all humankind. The physical zones of outer space do not belong to humankind—merely the freedom to access, explore, and/or use them. Article II of the OST expressly forbids states to appropriate them as national territories. Deep questions persist, such as the assumption in these phrases that space and celestial bodies are merely domains for humankind to explore, exploit, possess, and otherwise conquer and master. Perhaps different approaches to humankind's place and proper role are warranted (Eyres 2017). This avenue of thought crosses from space law, policy, and governance into broader cultural realms and needs further philosophical thinking across the social sciences (Figure 7.4).

7.9.3 Outer Space as a "Commons?"

One possible framework sometimes applied to some areas and some uses of space is the economic term "the commons," which has a long history of analysis in both legal and economic scholarship. There are both global commons and international commons regarding zones outside of state sovereignty. The Antarctic is an international commons shared between a select set of states in the Antarctic Treaty system, whereas the high seas are more globally accessed and utilized under the law of the sea. International law permitting peaceful universal access applies to outer space, as all states have the right or freedom to access outer space under the OST. Consequently, access to space might be considered as a global commons.

FIGURE 7.4 A size comparison between the planets Earth and Jupiter and the Moon. Are Jupiter, and the other planets, merely "resources" for humanity's use? (Adapted from Shutterstock, https://www.shutterstock.com/image-illustration/comparison-between-planets-earth-jupiter-moon-163046612.)

The access and use of useful orbits and useful frequencies are shared commons coordinated by the International Telecommunication Union (ITU). These frequencies and orbits are not owned by the states, but their use of particular orbits and frequencies is understood to be a right unilaterally enjoyed. However, calling space itself a global common is often overdone and misapplied, especially as space does not have a single sovereign ruler, is not comprised of a "territory," and is not a place where an economic foundation or situation innately requires collective use (such as a tract of arable land) (Hertzfeld et al. 2016).

Other newer labels include the phrase "common pool resources," and more legal terms like *res communis*, *res nullius*, and *res extra commercium* are also applied to space (Buck 1998). As mentioned at this chapter's outset, the striving for familiar phrases from academic fields and domains considered somehow similar to space means that these definitions are being applied to totally unique environments and activities far removed from their direct effects on Earth. We can rightly ask why should ancient concepts like *res communis*—built upon ancient Roman concepts of property—be applied to outer space? Should space activities force us to invent new understandings of rights, property, governance, and relations between actors—definitions based less on traditional concepts which must be adhered to, and more on the particular aspirations and particular concerns specific to the space domain and our evolving organizations and frameworks thereto? It seems likely that any worthwhile approaches will need to be flexible and inventive in developing new concepts and strategies.

7.10 SPACE SUSTAINABILITY BEGINS TODAY

Rational individuals are able to integrate long-term interests and values into the decision making in the present, which is why they seek to save for retirement, purchase insurance, and anticipate possible futures. In much the same fashion, actors in the space domain can take steps today which will serve them well in the long term. When we look at the current state of space governance including international laws, guidelines, standards, national licensing regimes, and even standard contracting practices of the aerospace industry we can ask whether these governance frameworks allow for a good future, a bad future, or something middling and in-between? How does our governance system balance current needs and desires with (both definite and predicted) future needs and desires?

7.11 CONCLUSION: THINK GLOBALLY, ACT LOCALLY

We are now in the second half of the first century of humankind's forays into outer space. It remains true that national governments are the major funders, managers, and (historically) the executors of space activities. Under existing international law, states remain the responsible entity for all national space activities. They bear the burdens of ensuring that national activities are in continuing compliance with international law, and of the potential liability for any physical damage caused.

Despite this, activities in space remain an inherently international endeavor, not just because it exists outside of state sovereignty, but because the projects and plans have international components here on Earth—with significant flows of capital, hardware, expertise, and personnel traveling where required to ensure the mission or project happens. Because space is globalized, it should be easier for stakeholders to realize that international coordination and cooperation is necessary, and that multilateral governance solutions have always been, currently are, and will remain necessary to sustainably manage and coordinate these activities.

The problem of "first movers" may crop up where a variety of actors must rely on each other for mutual benefit. For increasingly urgent issues such as space debris, why would one state be the first—or even an early—adopter of the types of stringent and costly governance regimes over its own activities while others are not so restrained? For naturally occurring hazards such as space weather and asteroid threats, why would states that rely less on space infrastructure, or with small populations or small areas of landmass, contribute as much as larger states with larger populations who perhaps rely more significantly on space infrastructure?

These problems are not unique to the space domain. But they are problems which must be faced squarely. Contemplating long-term interests for space sustainability is driven by the need for predictability by actors planning space activities. It is also driven by the need for a usable space environment, and the desire to save resources, which would otherwise be spent defending against hazards (including hazards from space weather and debris, but also significant resources saved by avoiding and preventing conflicts).

Other avenues are possible rather than arguing for global governance to address these shared concerns. International organizations such as the UN and related groups mentioned throughout this book, rather than imposing rules, can offer structure coordination and cooperation. The IADC, IAWN/SMPAG, COPUOS, ITU, and groups such as the Space Data Association all exist to pool knowledge and coordinate action. Their efforts are driven by the present need for space sustainability in its various guises. A sustainable future in space—helping to foster the future we want, and avoiding the future we fear—requires real efforts taken today. This involves factual findings about the space environment, imagining future scenarios (good and bad) and then comparing governance systems and the interests of all stakeholders in a discussion on whether current efforts either serve or fail to meet the needs of the future. Later, internalizing the norms of space sustainability, accepting the costs, and actually changing our actions are considered.

REFERENCES

Brundtland Commission, 1987. *Report of the World Commission on Environment and Development: Our Common Future*. United Nations. Retrieved from http://www.un-documents.net/our-common-future.pdf.

Buck S. J., 1998. *The Global Commons: An Introduction*. Washington, DC: Island Press.

Eyres H., 2017. *Seeing Our Planet Whole: A Cultural and Ethical View of Earth Observation*. Vienna: Springer.

Hertzfeld H., Weeden B., and Johnson C., 2016. Summer-Fall. Outer space: Ungoverned or lacking effective governance? New approaches to managing human activities in space (M. Cass-Anthony, Ed.). *The SAIS Review of International Affairs*, 36(2), 15–28.

Hitchens T., 2016. Forwarding multilateral governance of outer space activities: Next steps for the international community. In M. Simpson, R. Williamson and L. Morris (Eds.), *Space for the 21st Century* (pp. 75–117). Aerospace Technology Working Group.

Inter-Agency Space Debris Coordination Committee, 2007. *IADC Space Debris Mitigation Guidelines*. http://www.iadc-online.org/Documents/IADC-2002-01,%20IADC%20Space%20Debris%20Guidelines,%20Revision%201.pdf (accessed May 8, 2017).

Krag H., 2017. Mega challenges for mega-constellations. *ROOM—The Space Journal*, 1(11), 16–21.

Langewiesche W., 2004. *The Outlaw Sea: A World of Freedom, Chaos, and Crime*. North Point Press.

Martinez P., 2016. Sustainability, COPUOS and the LTSSA Working Group. In M. Simpson, R. Williamson, and L. Morris (Eds.), *Space for the 21st Century* (pp. 47–59). Aerospace Technology Working Group.

Secure World Foundation, 2014a. *Space Sustainability: A Practical Guide* (2014 ed.). Secure World Foundation. Retrieved from https://swfound.org/resource-library/brochures-and-booklets/.

Secure World Foundation, 2014b. *The UN Group of Governmental Experts on Space TCBMs*. Secure World Foundation. https://swfound.org/media/109311/swf_gge_on_space_tcbms_fact_sheet_april_2014.pdf (accessed May 8, 2017).

United Nations, 2013. *Group of Governmental Experts on Transparency and Group of Governmental Experts on Transparency and United Nations General Assembly*. New York: United Nations. http://www.un.org/ga/search/view_doc.asp?symbol=A/68/189 (accessed May 8, 2017).

United Nations Office for Outer Space Affairs, 2010. *Space Debris Mitigation Guidelines of the Committee on the Peaceful Uses of Outer Space*. Vienna: United Nations. Retrieved May 8, 2017, from http://www.unoosa.org/pdf/publications/st_space_49E.pdf.

CHAPTER 8

Space Activity and the Nascent Risk of Terrorism

Ben Middleton

CONTENTS

Despite the increased prevalence of terrorism and ubiquitous coverage of its devastating effects around the world, the probability of being killed in a terrorist attack is slight. It is axiomatic that at present there is only a very remote probability of activities in space being impacted by terrorism. This chapter provides a qualitative discussion of some of the terrorism risks, contextualized by factors relevant to quantitative risk analysis of terrorism (Willis et al. 2005). Quantitative studies typically categorize terrorism risk as a function of threat, vulnerability, and consequences. In this equation, the threat must comprise both the *intent* and the *capability* to perpetrate an attack; vulnerability must comprise an *assessment of the infrastructure to determine the probability of a damaging attack* occurring; and the consequences reflect an assessment of the *magnitude and type of damage* caused (Willis 2007). Due to an apparent lack of capability, the probability of a terrorist attack affecting space activity might be slight, but its potential impact severity is vast; a cursory analysis of this equation identifies potential lacunae in space law, policy, and regulation.

The framework of international law, first established by the Outer Space Treaty (OST), recognized that space exploration is to be carried out for the benefit and in the interests of all countries and is the "providence of all mankind". (Article I). Article III of the OST asserts the jurisdiction of international law in space "in the interest of maintaining international peace and security and promoting international co-operation and understanding." Article IV requires that states do not place weapons of mass destruction in space, and the Moon and other celestial bodies are to be used exclusively for peaceful purposes. Section 1(1) of the Draft International Code of Conduct on Space Activity states that its purpose is to "enhance the safety, security, and sustainability of all outer space activities pertaining to space objects, as well as the space environment." This international framework arose as a legal bulwark against the risks of the cold war era (Qizhi 1997), which does not reflect the reality of either the new commercial environment or the extant risks from military activities.

As the analysis of this chapter will show, the current risk from terrorism in space is likely to emanate from state-sponsored activity. Emerging space powers are entering the race and China's ambitious program of investment and development (Harvey 2013) adds a new Sino-axis to a chart that has traditionally depicted American (and formerly USSR) dominance. This will continue to influence the development of space policy and regulation and it is against this backdrop that the inexorable and largely unregulated drive toward commercialization of the space environment is occurring. It would be a mistake to ignore the risks to space activity posed by terrorism; this chapter maps some of these risks across the spectrum of probability, from real and present risks, through to more speculative, "blue-sky" threats. In examining these threats, the author examines ways in which a legal and regulatory framework might help to ameliorate some of the risks.

The analysis is split into three parts. First, the chapter examines the definition of terrorism in international law as it relates to space activity, before evaluating the existence of the threat in terms of terrorist intent and capability. Second, the discussion assesses the vulnerability of space activity to terrorism, together with the potential consequences of such an attack. Threats posed by nuclear terrorism, threats to satellites and the impact on communications, remote sensing, and global navigation satellite systems (GNSS) are considered, alongside a discussion of the risk terrorism could pose to the embryonic space tourism industry. Finally, the chapter identifies that the reality of the terrorism threat in space is that it is currently a state-based phenomenon, which is difficult to directly address through International law. Instead, the chapter suggests ways in which commercial operators and state actors need to respond in order to minimize the risk from terrorism as the capability of terrorist groups increases. The chapter concludes that international cooperation is needed to secure an operational regulatory framework, which is supported by commercial incentives for private enterprise.

8.1 DEFINING TERRORISM IN INTERNATIONAL LAW

Terrorism is quintessentially a human phenomenon (Hoffman 2006); extending the reach of human activity beyond our planet may have the unfortunate consequence of extending the tendrils of terrorism into space. In order for an activity to be prohibited, however, it first must be properly defined. Despite the events of 9/11, which represented a paradigm shift in the nature of the terrorism threat, the international community has struggled to agree to a single unified definition of terrorism (Hoffman 2006). In their respected work, Schmid and Jongman (1987) suggest that terrorism is:

> an anxiety-inspiring method of repeated violent action, employed by (semi-) clandestine individual, group or state actors, for idiosyncratic, criminal or political reasons, whereby—in contrast to assassination—the direct targets of violence are not the main targets.

There have been numerous United Nations conventions on terrorism to cover different specified acts of terrorism, although to date UN definitions have eschewed consideration of political motivation (Byrnes 2002). Sidestepping the thorny issue of the definition, UN resolution 1373 (2001) secured an agreement on prevention and suppression of terrorist financing, declaring that states shall "refrain from providing" any support to terrorists, take steps to prevent terrorism, and bring terrorists to justice. Paragraphs 3 and 6 of this resolution required cooperation between states and established the Counter-Terrorism Committee to oversee implementation. The adoption of the United Nations Global Counter-Terrorism Strategy (2006, A/RES/60/288) could have provided the impetus for agreement on a definition, but despite the strategy's description of terrorism as "one of the most serious threats to international peace and security" and urging of the strengthening of international cooperation and the implementation of General Assembly and Security Council resolutions, UN attempts to agree on the scope of a definition and exceptions to it have stalled. The aphorism "one man's terrorist is another man's freedom fighter" explains why further immediate progress is unlikely: disagreements center on a distinction between terrorism and the legitimate right of people to resist foreign occupation (Walter 2003). There is also inevitable disagreement as to the existence of liability for state-sponsored terrorism (*Ibid*).

8.1.1 State-Sponsored Terrorism and the Militarization of Space

There are well-documented lacunae in the international legal framework when it comes to militarization of space-based activity (see Markoff 1976) and there is potential for crossover when it comes to consideration of terrorism in space. A state's military and scientific prowess, and economic might, can be ably demonstrated by the existence of an ambitious space program. North Korea's missile development activities operate under the guise of its space program (Sang-Hun 2012). Anti-Satellite (ASAT) missiles or "kinetic kill" vehicles have been used by both the United States and China (Weedon 2014). As Remuss (2009) notes, the real risks of international terrorism in outer space are currently state-based, and policy considerations appear to adopt this as their focus. Indeed, Jasani (2016) has suggested that there is the need for an international treaty to prohibit anti-satellite weapons in order to reduce the risk to the space environment. Given the complexities with securing international agreement on space-based activity,* it is suggested that, for now at least, any attempt to formulate a unified definition of terrorism, which could apply to space-based activity, is probably doomed to failure. Instead, broad principles, which feature within common definitions, can be used in order to inform the evolution of law, policy, and regulation in space.

---

* See, for example, the fact that none of the main spacefaring nations have ratified the Agreement Governing the Activities of States on the Moon and Other Celestial Bodies: Marko (1993).

8.1.2 Possessing Terrorist Intent

Consideration of terrorism in space should include an analysis of the motivations and intent behind any potential attack, since it is directly relevant to the existence of a threat. A common element of most terrestrial definitions of terrorism is the need for a sense of intimidation, fear, or panic in the populace (Schmid 2011). The distinction is important here for a number of reasons. In some instances it might help to differentiate state-based terrorism in space from threats by a nongovernment actor (the aims of the former might not require such intimidation). It also could impact on the probability of the manifestation of such a threat. For example, if a terrorist group disabled a satellite, it is more likely that this would cause economic damage and inconvenience to the public, rather than generate the climate of fear and anxiety upon which the motivations of terrorist groups depend. Regrettably, on Earth there is no shortage of soft targets, attacks on which require comparatively little effort or resource to perpetrate, as has been tragically demonstrated in recent years. Even in the absence of intimidation or fear, the deliberate targeting of space-based activity, for example through cyber attack or explosion, could still be designed to influence the government of a state, and such activity would still befit a terrorism label.

Definitions of terrorism often include the use or threat of violence (Hoffman 2006; Schmid 2011) and it is suggested that this is too narrow to encapsulate emerging threats to space-based activity. Threats to space-based infrastructure, or to infrastructure on the ground which supports space activity, do not necessarily have to be violent in the traditional sense of the word in order to lead to damaging consequences. Just as the terrestrial threat of terrorism has shifted (Global Terrorism Index 2015), the threat from terrorism in space activity clearly includes the threat of cyber attack, and this is one area where terrorists could feasibly possess the capability to exploit vulnerability in order to further their cause. In this context, the UK definition of terrorism is helpful, since it expressly includes action "designed seriously to interfere with or disrupt an electronic system" (s. 1(2)(e) Terrorism Act 2000; Legislation.gov.uk, n.d.). While imagining the impact of terrorism in space might not immediately conjure up an image of this form of attack, it is arguably the most likely in the short term.

8.1.3 Capability of Terrorist Groups

It is possible that in the future a terrorist group could be motivated to carry out an attack which impacts on activities in space, perhaps by an attack on multiple satellites in low Earth orbit (LEO), or an attack on the space tourism industry, and even, in the most extreme scenario, by a nuclear explosion. When the intent of terrorist groups to perpetrate an act of terrorism in space is considered against their capability to do so, however, the probability of the threat becomes *de minimis*. Capability might be compromised on the grounds of financial or physical resources, a lack of technical expertise, or both. Certainly the likelihood that a non-state supported terrorist group possesses the current capability to carry out an attack impacting on space activity is remote. In the wake of 9/11, the United Nations has achieved some success in disrupting the sophisticated financial resource networks of terrorist organizations through UN Resolution 1373 (Bianchi 2006). Coordinated international activity has made it much more difficult for terrorist groups to finance their operations, and indeed forensic accounting is now used an investigative tool when it comes to disrupting future terrorism activities (Biersteker and Eckert 2008). Hypothetically, some terrorist groups may retain the infrastructure and available financial resource in order to consider such an attack, but at present it is unlikely that they will be motivated to do so in light of the scale of the technological challenge required.

Of the most probable outcomes, a cyber attack designed to disrupt the operation of satellites is a real threat. A terrorist group has already used the space environment to pursue its ideology: in 2007, the Liberation Tigers of Tamil Eelam (LTTE), known as the Tamil Tigers and designated as a terrorist organization across multiple states, hijacked a U.S.-based satellite transponder and used it to broadcast their terrorist propaganda across India (Al-Rodhan 2012). The armed forces of Sri Lanka have since defeated the LTTE, but the first use of satellite technology in this way by such a well connected and resourced terrorist group tells a cautionary tale (Dharmawardhane 2016). True, such an operation would be complex and costly for a terrorist group to attempt, but the pace of developments in the space industry should highlight this as a genuine risk. Capabilities of private enterprise are expanding rapidly: even the most conservative predictions would suggest an inexorable expansion of space activities over the next decade. Additionally, while not all space activities are currently vulnerable to the threat posed by terrorism, the magnitude of the potential consequences is vast. In the coming years, the threat from terrorism in space could shift from *de minimis* to *de manifestis,* and an international response in terms of law and policy is required.

8.1.4 Vulnerability to Nuclear Terrorism

One of the worst-case scenarios, the detonation of a nuclear warhead in the atmosphere, is clearly an extremely remote possibility in the medium term yet still has the potential to be devastating. An indication of the severity of the potential implications was provided in the notorious U.S. test "Starfish Prime," a nuclear detonation which unexpectedly caused a belt of radiation around the earth, crippling numerous satellites and causing electrical interference through a much larger than anticipated electro-magnetic pulse (Hollingham 2015). More tests followed and the events of the Cuban Missile Crisis then led to the signing of the Atmospheric Test Ban Treaty (Dupont 2004). Although there have not since been nuclear detonations in space, the cautionary lessons of Starfish Prime should not go unheeded. There are several hundred satellites in LEO which are not shielded against the kind of radiation that a nuclear detonation at an altitude of between 100–300 km would cause. It has been predicted that 90% of satellites in LEO could be disabled within a month if a small nuclear warhead, mounted on a ballistic missile, was detonated at this altitude, and many states have the ability to carry out such an attack (Dupont 2004). Although the majority of military satellites have been designed to be shielded from the effects of this radiation, and many are positioned at higher altitudes such that they are unlikely to be affected, the consequences would still be enormous, not least to commercial satellites, upon which numerous state armed forces still depend. Such an attack could require manned space activity to be paused, and the cost of replacement satellites, to say nothing of the associated economic impact on Earth, would be colossal.

Outside the threat of nuclear detonation in orbit, Kuperman (2013) notes that a further area of potential risk arises from satellites powered by Highly Enriched Uranium (HEU) reactors, which have been used by Russia and the United States. Although their use has waned in recent years, some of these satellites remain in orbit (Messer 2013). If current or emerging space powers elect to revisit the use of HEU in their space programs, it is likely that security of the material during transport and the safety of such material during launch represent the greatest risks (Potter 2010:20).

The threat from nuclear terrorism in space should be taken seriously, but it should be recognized that for now at least an attack is unlikely to be perpetrated by a nonstate actor. A terrorist group would be able to cause far worse damage and loss of life, and create significantly more fear if it managed to detonate a device in a populated area on Earth (Lewis

2017). As a nuclear state with few diplomatic relationships, North Korea probably poses the greatest immediate risk, and given the response of the international community to its activities it is unlikely that this risk can be ameliorated by further regulation alone: states must continue to work together to stymie Pyongyang's ability to further develop its ballistic missile and nuclear technologies.

8.1.5 Vulnerability of Satellites

In the short term, the risk from terrorism chiefly concerns targeted activity taken against satellites, which in turn would degrade states' communications and ability to conduct remote sensing, reconnaissance, and missile defense (Remuss 2009). There are more than 1400 satellites currently in orbit, with more than 800 in LEO and more than 500 in geostationary orbit (GEO) (UCS Satellite Database, 2017). As Livingstone and Lewis (2016) note, the positioning of satellites in LEO renders them particularly vulnerable to hostile activities of states; amateur astronomers can readily observe satellites in LEO and the ability to track a satellite makes it easier to attack (Remuss 2009).

The most obvious current risk to satellites in LEO comes from ground-based jamming—broadcasting a higher energy signal at the same frequency at a satellite in order to drown out the signal, usually to overwhelm terrestrial receivers (Livingstone and Lewis 2016). Though illegal in many countries, GPS jammers, which operate within a small radius are readily available on the Internet (Rügamer and Fraunhofer 2015). GNSS signals are not powerful and it does not take a particularly large jamming device to cause disruption on a localized scale. Hayden (2016) observes that there is substantial evidence that this technology has been used against U.S. forces in Afghanistan and Iraq. However, military satellites are equipped with anti-jam technology, while GNSS jamming detectors have been developed together with guided missiles sensitive enough to destroy jamming equipment (Strategypage 2016). As such the threat from jamming is likely to have a greater impact on unprotected commercial applications.

More hypothetically, it may be possible for an Earth-based use of a high energy LASER to "blind" satellites in LEO, medium Earth orbit (MEO), and even beyond to GEO. This could cause degradation of the GNSS service or disruption to communications (Remuss 2009). However, it is suggested that the technology required to perpetrate such an attack, its chances of success, and its potential impact are such that this must be regarded as only an extremely remote possibility, and one which again remains the preserve of state actors. Physical attacks on satellites in MEO and GEO are also less likely.

In terms of the probability of risk, a ground-based cyber attack is much more significant: a well-resourced terrorist group could have the potential to carry out a cyber attack on a satellite (Livingstone and Lewis 2016). There is also a significant risk posed by individual hackers who want to demonstrate their prowess and highlight vulnerability, even in the absence of terrorism-related intent (Fritz 2012). There have been several reports of successful hacking attacks (USCC 2011:8), including in 2014 when the U.S. National Oceanographic and Atmospheric Administration reported that its satellite was taken offline in 2014 as a result of actions by hackers, reportedly based in China (Flaherty et al. 2014). Commercial satellites may be particularly vulnerable when they rely on over the air software updates, which are broadcast from ground-based systems that could be subject to cyber attack (Remuss 2009). Ground stations themselves are vulnerable to physical attacks which could render satellites inoperable.

8.2 POTENTIAL CONSEQUENCES OF AN ATTACK ON SATELLITES

An attack on one or several satellites could have a myriad of potentially severe consequences. Disruption to civilian and military communications is possible: although U.S. military communications are dependent on satellites in GEO and are generally believed to be resistant to outside interference the bulk of military communications are still routed through unsecured commercial satellites, which represent an Achilles' heel (Chun 2012). Disruption to these satellites could cause problems with communications networks on Earth, with notable economic impact. However, the magnitude of the risk in terms of military communications is continuing to reduce with the introduction of new systems, many with multiple redundancies. The Advanced Extremely High Frequency (AEHF) system is the product of a joint venture between the United States, the United Kingdom, Canada, and the Netherlands, and has the potential to be able to provide topography and maps of battlefield positions and live video for military purposes in order to "more quickly and safely command some of the world's most capable militaries to address global instability as it arises" (Lockheed Martin 2015), as well as providing a secure communication network resistant to jamming. Simultaneously, a new system of military satellites in GEO, the Wideband Global Satcom (WGS) network, is currently being deployed with completion expected during 2018 (Wall 2016).

The sheer number of satellites in LEO with remote sensing capabilities means there are multiple inbuilt redundancies, in addition to the fact that states can also depend on satellite technology of their allies. Therefore, unless a terrorist attack affected large swathes of LEO it is unlikely that there would be complete loss of remote sensing ability. However, a disruptive attack on a satellite in LEO could result in degradation of this service. This could have an impact on military operations, including an impact on the fight against terrorism itself: Lee and Steel (2014) document the military use of satellite applications, particularly in relation to the U.S.-led invasion of Afghanistan and Iraq following the 9/11 World Trade Center attacks. Imagery is provided by both military satellites and commercial operators; the latter license their product up to permitted resolutions greater than 25 cm. Morgan (1994) has explored the significant military use of commercial satellites, and this reliance exposes vulnerability since these satellites rarely possess anti-jam technology or hardening against the threat of radiation. The precise resolution of modern military remote sensing satellites is classified, but the technology is likely to be superior both in terms of resolution and the ability to cover more of the electromagnetic spectrum. Such imagery has proven itself invaluable to identify high profile targets and to support ground operations; disruption to it would prove costly.

The consequences of greatest magnitude, however, are likely to come in the event a satellite is hijacked and control over some of its systems is gained, or if it is remotely disabled. At the low end of the probability spectrum, if a hijacked satellite is equipped with its own propulsion system, it could be deliberately used as a weapon (notwithstanding the formidable technical difficulties involved in such an operation). Destruction of a satellite or collision in space could create a debris field, which theoretically could precipitate the Kessler syndrome (Kessler et al. 2010) with catastrophic consequences both in space and on Earth.

If a satellite's systems were compromised, a more probable risk would be the transmission of a false or unwanted signal. This phenomenon is known as "spoofing" and is not new (Psiakim and Humphreys 2016). Feeding disinformation to GNSS receivers on the ground could mean that vehicles or vessels are deliberately led astray, with potentially devastating consequences; proof-of-concept tests have shown that this is possible (UTNews 2013). From the perspective of everyday life and military capability, perhaps the most obvious impact of a terrorist attack on satellites would be one which substantially degraded GNSS capability. The use of

GNSS technology is ubiquitous, providing real time navigation data which is relied on, *inter alia,* by armed forces and emergency services the world over. It has been entrenched in the mining and agriculture industries, as well as its more obvious aviation, maritime, and increasingly civilian recreational uses. GPS provides a precise time stamp which is relied on by telecommunications networks to synchronize communications, financial institutions to time their transactions, and even power stations to regulate their systems (Sadlier et al. 2017). Its usage is so pervasive it could easily be considered part of the critical infrastructure of a state (National PNT Advisory Board 2010); the position in the United States is currently to categorize it as an “enabler” for the critical infrastructure (Divis 2015).

An attack on GNSS would probably not take the form of destruction of an individual satellite, which would be a costly operation unlikely to significantly degrade GNSS functionality. GNSS technology now includes the GPS, GLONASS, Galileo, Beidou, and other systems each of which have multiple redundancies. However, the risk should not be underestimated. In 2016, during the decommissioning of an U.S. Air Force satellite, a software issue resulted in 15 GPS satellites broadcasting the incorrect time by 13 microseconds (Baraniuk 2016). The error was quickly corrected, and backup systems ensured that disruption was minimized, but telecommunications networks encountered 12 hours of errors and some digital radio services suffered interference (Baraniuk 2016). Loss of GNSS service even for a short time would have a huge economic impact. In the United Kingdom, for example, Sadlier et al (2017) have estimated that sectors generating 11.3% of GDP are reliant on GNSS technology to some extent, and 5 days of disruption to the service are predicted to cost the UK economy £5.2bn. Degradation of GNSS service over an extended period could prove even more costly, as inaccurate information could cause an effect similar to spoofing (Sadlier et al. 2017). The Royal Academy of Engineering (2011) also implies that disruption to GNSS could pose a significant risk to human lives.

8.3 HORIZON SCANNING: FUTURE SPACE-BASED TERRORISM

Elon Musk recently announced that SpaceX will offer two places for private citizens to the Moon (Malik 2017). Russia has stated that Soyuz spacecraft could be used for dedicated space tourism missions, potentially taking two tourists and one cosmonaut to the ISS (Tass 2017). Virgin Galactic is likely to be the first to open up space tourism to the (comparatively) mass market (Klotz 2017). It is possible that within 10–20 years, civilian access to space could be priced in the tens of thousands of dollars. This shift will have profound consequences, and is likely to present ongoing challenges to the legal and regulatory framework in space (Newman 2016). In terms of the risk from terrorism, space tourism missions are likely to be most at risk during launch and reentry; comprehensive screening processes will be necessary for passengers and the chance of a passenger perpetrating an attack midflight would therefore be slim. Ground-based security measures at spaceports will be necessary to protect the physical environment of the launch. Suborbital flights would present a similar risk.

Long-term space missions, such as a potential human mission to Mars, would need to be considered although a much more substantial framework to cover conduct and behavior on such missions will be necessary well before terrorism poses a risk (Newman 2016) and psychological screening and background checks are likely to be formidable. In terms of impact, a terrorist attack on a space vehicle carrying tourists clearly has the potential to cause mass loss of life. Just as concerning, however, would be an attack—perhaps sabotage of a critical system—which could precipitate a drawn out crisis. In such a situation it might be necessary to starve the terrorists of the oxygen of publicity, in order to deprive them of the atmosphere of intimidation and fear upon which they depend.

8.4 CONCLUSIONS: GUIDING PRINCIPLES FOR INTERNATIONAL SOLUTIONS

As space activity continues to expand, the nascent risk terrorism poses cannot be ignored: the *de minimis* risk could eventually manifest itself with devastating consequences. The question of how a framework of law and policy respond to such a threat must be considered. Work to minimize the risk must inevitably be contingent on further technological developments designed to minimize the potential consequences. While this approach is advocated in the context of fighting terrorism on the final frontier, its benefits could be pervasive across multiple industries. The hardening of future sensitive commercial satellites in LEO, and increasing capacity so as to increase the redundancies in the system would be a good start. The ability to rapidly replace a critical satellite could be explored. Increasing codependency in satellite systems where states work together to share services will undoubtedly continue, but from the perspective of risk this should be considered a double-edged sword if it has the potential to broaden the impact of the consequences. Individual states should be encouraged to designate GNSS technology as part of their critical infrastructure and conduct more detailed assessments of its vulnerability and risk. Experience with satellite jamming and anti-jamming indicates that sometimes technology can provide both the problem and the solution. Beyond that, however, changes to the regulatory framework are undoubtedly required.

8.4.1 Legal Responses

Space is not free from law, nor from the rule of law (Wessel 2012); international treaties and nonbinding agreements provide rudimentary legal foundations upon which we must continue to build. The importance of this approach should not be underestimated: a strong framework of international law, which underpins respect for the rule of law, is anathema to the concept of terrorism. There can be no compelling argument made to increase the weaponization of space in order to prevent terrorism: this would be self-defeating and would instead increase the risk to the international order. Maintaining the principle of peaceful exploration is essential, and can actually serve to prevent terrorism: satellites provide an invaluable mechanism of directly fighting terrorism through quasi military uses such as remote sensing, GNSS, missile defense, and communications. Indeed, in 2016, Egypt and Syria attempted to add an agenda item to COPUOS in order to encourage the sharing of detailed remote sensing imagery between states in the fight against terrorism (COPUOS 2016), although this was unsuccessful. More broadly, continuous peaceful exploration and the emerging space tourism industry will expose the human race to reality of the fragility of life on our planet: this "overview effect" can only have a positive impact on the resolution of our political and ideological divides (White 1998).

Progress on an international definition of terrorism which applies to space activity is unlikely: there has not yet been such agreement on Earth and a large element of the risk of terrorism emanates from state-sponsored threats. As Remuss (2009) suggests, a policy aimed at countering terrorism in space might provide an ultimate solution, although this would appear to be ambitious and unrealistic at present. Until this can be achieved, initial guiding principles can be considered.

8.4.2 Space Policy and Regulation

Technology exists in order to secure satellites against jamming and other forms of cyber attacks, but the next few decades will see an increased commercialization of the space environment, and in that environment market forces will determine how satellites are designed and developed. The prohibitive cost of "hardening" civilian satellites in order to make them resistant to radiation has meant that commercial enterprises will not accede to it voluntarily (Dupont 2004). There is a risk that without regulation a proliferation of unsecure satellites, which could be vulnerable to outside interference or attack, could have damaging implications for the space environment in LEO for all users. It is unlikely that there is sufficient impetus for wholesale regulation, and it is important that regulation does not unduly impede commercial investment and development. Continuous steps are needed toward an international framework which has the support of, but is not unduly dominated by, current state actors.

Livingstone et al. (2016:15) have suggested that one way to achieve this would be to embed a culture of cybersecurity in the commercial supply chain, with a default position being the provision of a "lightly regulated" compliance framework which does not stifle commercial innovation but rather will allow business interests to become the principal driver of cyber security. There is merit in this multistakeholder space cybersecurity approach (*op cit* 37) and it continues a trend toward "soft law"—politically binding arrangements in which nonstate actors can play a role (Tannenwald 2004). Commercial sector buy-in could be encouraged by incentives, *inter alia* to ensure the protection of critical commercial satellites or dual-use satellites (Remuss 2009). There is a need to more fully develop the broad rhetorical principles laid down by the OST into something which can be used by space actors in a new, commercialized environment. As Tannenwald (2004:412) argues, "the rules of space will need to reflect a global, rather than national, public interest, and not merely the interests of a few spacefaring governments and corporations." At a state level, COPUOUS might provide a forum in which these issues can be discussed; there needs to be in place a regime which incorporates "a strong normative prohibition" (*Ibid* 416) against interference with space assets.

It is unlikely that the lacunae in law and policy related to terrorism activity in space will be quickly filled. Terrorist capability in space is currently limited: while the motivation might exist, in the absence of state support terrorist organizations are unlikely to possess the resources or technical expertise required to perpetrate an attack. However, this is unlikely to remain the case indefinitely and there are vulnerabilities in space infrastructure, which can potentially be exploited by state actors and nonstate supported terrorists alike. Given the magnitude of the potential consequences there is significant risk in continued inaction. As was stated in another context by the

9/11 Commission Report (2004: 398): "[p]rivate-sector preparedness is not a luxury; it is a cost of doing business in the post-9/11 world. It is ignored at a tremendous potential cost in lives, money, and national security."

REFERENCES

9/11 Commission Report, 2004. https://www.9-11commission.gov/report/911Report.pdf.

Al-Rodhan, N., 2012. Space security and meta-geopolitics. In *Meta-Geopolitics of Outer Space: An Analysis of Space Power, 95*, 69–100, Palgrave Macmillan.

Baraniuk, C., 2016. "GPS error Caused '12 Hours of Problems' for Companies" *BBC News*, http://www.bbc.co.uk/news/technology-35491962, February 4, 2016.

Bianchi, A., 2006. Assessing the effectiveness of the UN Security Council's anti-terrorism measures: The quest for legitimacy and cohesion. *Eur. J. Int. L.*, 17(5): 881–919.

Biersteker, T. and Eckert, S., 2008. *Countering the Financing of Terrorism*. Oxford: Routledge.

Byrnes, A., 2002. "Apolcalyptic Visions and the Law: The Legacy of September 11," *Inaugural Lecture*, https://openresearch-repository.anu.edu.au/bitstream/1885/41104/3/Byrnes30May02.pdf.

Chun, C., 2012. *Defending Space: US Anti-Satellite Warfare and Space Weaponry.* Osprey.

COPUOS, 2016. Committee on the Peaceful Uses of Outer Space, Combating Terrorism Using Space Technology: Proposal to Add a New Agenda Item in the Year 2017, http://www.unoosa.org/res/oosadoc/data/documents/2016/aac_1052016crp/aac_1052016crp_18_0_html/AC105_2016_CRP18E.pdf.

Dharmawardhane, I., 2016. "Sri Lanka." In *Handbook of Terrorism in the Asia Pacific*. eds R. Gunaratna and S. Kam, 2016. 320–321. Imperial College Press.

Divis, D.A., 2015. "PNT Advisory Board Debates Critical Infrastructure Designation for GPS," *Inside GSS*, http://www.insidegnss.com/node/4536. June 16, 2015.

Dupont, D., 2004. Nuclear explosions in orbit. *Sci. Am.*, 290, 100–107.

Flaherty, M., Samenow, J., and Rein, L., 2014. "Chinese Hack U.S. Weather Systems, Satellite Network," *Washington Post*, November 12, 2014, https://www.washingtonpost.com/local/chinese-hack-us-weather-systems-satellite-network/2014/11/12/bef1206a-68e9-11e4-b053-65cea7903f2e_story.html?utm_term=.35c8d30db84a.

Fritz, J., 2012. "Satellite Hacking: A Guide for the Perplexed" Culture Mandala: Bulletin of the Centre for East-West Cultural and Economic Studies, vol. 10, No. 1, December 2012–May 2013, pp. 21–50, http://www.international-relations.com/CM2012/Satellite-Hacking.pdf.

Global Terrorism Index, 2015. *Measuring and Understanding the Impact of Terrorism*. Institute for Economics and Peace, http://economicsandpeace.org/wp-content/uploads/2015/11/Global-Terrorism-Index-2015.pdf.

Harvey, B., 2013. *China in Space*. Springer Praxis.

Hayden, M.V., 2016. *Playing to the Edge: American Intelligence in the Age of Terror.* New York: Penguin.
Hoffman, B., 2006. *Inside Terrorism*, 2nd ed. Chichester: Columbia University Press.
Hollingham, R., 2015. *BBC*, http://www.bbc.com/future/story/20150910-the-nuke-that-fried-satellites-with-terrifying-results.
Jasani, B., 2016. "Space Assets and Emerging Threats," http://www.unoosa.org/pdf/SLW2016/Panel2/1_Jasani_-_Space_assets_and_threats_06082016.pdf.
Kessler, D.J., Johnson, N.L., Liou, J.C., and Matney, M., 2010. The Kessler syndrome: Implications to future space operations. *Adv. Aeronaut. Sci.*, 137, 8.
Klotz, I., 2017. "Virgin Galactic Aims to Fly Space Tourists in 2018, CEO Says," https://www.space.com/36654-virgin-galactic-fly-space-tourists-2018.html.
Kuperman, A.J., 2013. Global HEU phase-out: Prospects and challenges. In *Nuclear Terrorism and Global Security: The Challenge of Phasing our Highly Enriched Uranium, 19.* Routledge.
Lee, R. and Steel, S., 2014. Military use of satellite communications, remote sensing, and global positioning systems in the war on terror. *J. Air Law Commer.*, 79, 75.
Lewis, J., 2017. "Would a North Korean Space Nuke Really Lay Waste to the US?" *New Scientist*, May 3, 2017, https://www.newscientist.com/article/2129618-would-a-north-korean-space-nuke-really-lay-waste-to-the-us/.
Livingstone, D. and Lewis, P., 2016. Space: The Final Frontier for Cybersecurity. Chatham House Research Paper, September 2016, https://www.chathamhouse.org/sites/files/chathamhouse/publications/research/2016-09-22-space-final-frontier-cybersecurity-livingstone-lewis.pdf.
Malik, T., 2017. "As SpaceX Unveils Space Tourist Moon Flight, NASA Reacts" February 28, 2017, https://www.space.com/35850-spacex-private-moon-flight-nasa-reaction.html.
Markoff, M.G., 1976. Disarmement and "peaceful purposes" provisions in the 1967 Outer Space Treaty. *J. Space L.*, 4, 3–22.
Marko, D.E., 1993. A kinder, gentler moon treaty: A critical review of the current moon treaty and a proposed alternative. *J. Nat. Resour. Envtl. L.*, 8, 293–346.
Martin, L., 2015. Press Release, "Protected Military SatCom Network Live for Global Deployment" July 31, 2015, http://www.lockheedmartin.co.uk/us/news/press-releases/2015/july/space-AEHF-IOC.html.
Messer, R.B., 2013. Space reactors. In *Nuclear Terrorism and Global Security: The Challenge of Phasing out Highly Enriched Uranium*. ed. A.J. Kuperman, 211–213. Oxford: Routledge.
Morgan, R.A., 1994. Military use of commercial communication satellites: A new look at the outer space treaty and peaceful purposes. *J. Air L. Com.*, 60, 237.
National PNT Advisory Board, 2010. Comments on Jamming the Global Positioning System—A National Security Threat: Recent Events and Potential Cures, November 4, 2010 http://www.gps.gov/governance/advisory/recommendations/2010-11-jammingwhitepaper.pdf.
Newman, C.J., 2016. The way to Eden: Environmental legal and ethical values in interplanetary space flight. In *The Ethics of Space Exploration*. eds. J.S.J. Schwartz and T. Milligan, 221–238. Springer.

Potter, W., 2010. Nuclear terrorism and the global politics of civilian HEU elimination. In *The Global Politics of Combating Nuclear Terrorism: A Supply-Side Approach*. eds. W. Potter and C. Hansell, 20–21. Oxford: Routledge.

Psiakim, M.L. and Humphreys, T.E., 2016. "GNSS Spoofing and Detection," *Proceedings of the IEEE*, https://radionavlab.ae.utexas.edu/images/stories/files/papers/gnss_spoofing_detection.pdf.

Qizhi, H., 1997. The Outer Space Treaty in perspective. *J. Space L.*, 25(2), 93.

Remuss, N., 2009. *The Need to Counter Space Terrorism—A European Perspective*. European Space Policy Institute, https://www.files.ethz.ch/isn/124638/espi%20perspectives%2017.pdf.

Royal Academy of Engineering, 2011. Global Navigation Space Systems: Reliance and Vulnerabilities, http://www.raeng.org.uk/publications/reports/global-navigation-space-systems.

Rügamer, A. and Fraunhofer, I.I.S., 2015. "Jamming and Spoofing of GNSS Signals—An Underestimated Risk?!" https://www.fig.net/resources/proceedings/fig_proceedings/fig2015/papers/ts05 g/TS05G_ruegamer_kowalewski_7486.pdf.

Sadlier, G., Flytkjær, R., Sabri, F., and Herr, D., 2017. *The Economic Impact on the UK of a Disruption to GNSS: Full Report, London Economics*, https://www.gov.uk/government/uploads/system/uploads/attachment_data/file/619544/17.3254_Economic_impact_to_UK_of_a_disruption_to_GNSS_-_Full_Report.pdf.

Sang-Hun, C., 2012. "North Korean missile said to have military purpose," *New York Times*, December 23, 2012, http://www.nytimes.com/2012/12/24/world/asia/north-korean-rocket-had-military-purpose-seoul-says.html?_r=1&.

Schmid, A. and Jongman, A., 1987. *Political Terrorism*. North-Holland, Amsterdam: North Holland Publishing.

Schmid, A. 2011. *The Routledge Handbook of Terrorism Research*. Oxford: Routledge.

Strategypage, 2016. "Electronic Weapons: The Antidote For GPS Jamming," https://www.strategypage.com/htmw/htecm/20160921.aspx.

Tannenwald, N., 2004. Law versus power on the high frontier: The case for a rule-based regime for outer space. *Yale J. Int'l L.*, 29, 363.

Legislation.gov.uk. (n.d.). Terrorism Act 2000. http://www.legislation.gov.uk/ukpga/2000/11/contents

Tass, 2017. 'Russia Ready to Send Tourists into Space' June 20, 2017. http://tass.com/science/952384.

UCS Satellite Database, 2017. Union of Concerned Scientists, http://www.ucsusa.org/nuclear-weapons/space-weapons/satellite-database#.WVAIdcaZNPMht

United Nations Security Council Resolution (UN resolution) 1373, [threats to international peace and security caused by terrorist acts], November 12, 2001, S/RES/1377.

US- China Economic and Security Commission (USCC), 2011. *Report to Congress of the US-China Economic and Security Commission*, https://www.uscc.gov/sites/default/files/annual_reports/annual_report_full_11.pdf.

UTNews, 2013. University of Texas at Austin, 'UT Austin Researchers Successfully Spoof an $80 million Yacht at Sea' https://news.utexas.edu/2013/07/29/ut-austin-researchers-successfully-spoof-an-80-million-yacht-at-sea.

Wall, M., 2016. "U.S. Air Force Launches Advanced Military Communications Satellite" December 2016. http://www.space.com/34950-air-force-launches-wgs8-military-satellite.html.

Walter, C., 2003. "Defining Terrorism in national and International law," http://iusgentium.ufsc.br/wp-content/uploads/2017/03/1-2-Defining-Terrorism-in-National-and-International-Law-Christian-Walter.pdf.

Weedon, B., 2014. *Through a Glass, Darkly Chinese, American, and Russian Anti-satellite Testing in Space*. Secure World Foundation, https://swfound.org/media/167224/through_a_glass_darkly_march2014.pdf.

Wessel, B., 2012. The rule of law in outer space: The effects of treaties and nonbinding agreements on international space law. *Hastings Int'l Comp. L. Rev.*, 35, 289 2012, 322. Reston: American Institute of Aeronautics and Astronautics, Inc.

White, F., 1998. *The Overview Effect: Space Exploration and Human Evolution (Library of Flight)*, 2nd edn.

Willis, H.H., Morral, A.R., Kelly, T. K., and Medby, J., 2005. *Estimating Terrorism Risk*. MG-388-RC. Santa Monica, CA: RAND Corporation.

Willis, H.H., 2007. *Guiding Resource Allocations Based on Terrorism Risk*. Published Articles & Papers. Paper 28. https://research.create.usc.edu/cgi/viewcontent.cgi?referer=http://scholar.google.co.uk/&httpsredir=1&article=1062&context=published_papers.

CHAPTER 9

Reconciling the Past, Present, and Future of National Security, Military Activity, and Space Law

Michael J. Listner

CONTENTS

9.1 INTRODUCTION

With regard to precipitous heights, if you proceed your adversary, occupy the raised and sunny spots, and there wait for him to come up.

Remember, if the enemy has occupied precipitous heights before you, do not follow him, but retreat and try to entice him away.

SUN TZU
(Clavell, pp. 51–52)

These words from a general from the Chinese Kingdom of Wu around 510 AD were couched in terms of warfare at the time, but their meaning has transcended the ages and is relevant to the present and takes on a new perspective with regard to outer space and the role it plays for the national security of many nations, especially the United States. The emphasis of outer space on national security was apparent even before the launch of *Sputnik-1*, but the questions of how the geopolitical environment would respond were many. Post-*Sputnik*, some of the questions of how the legal environment would evolve were taken into consideration not just from the perspective of civilian uses but for national security.

Many of the early thoughts on the uses of outer space included the perspective from military planners and manifested themselves in many projects—most of which never reached fruition. Indeed, it could be said outer space was looked upon as a theater for future warfare; however, the role of the military changed with the signing of the Outer Space Treaty (OST), adding a new facet to the jurisprudence of international law which presented space as a global common of sorts. The OST was a reaction to the burgeoning potential for outer space to be appropriated by a large power and was the second of the "nonarmament" treaties (Department of State 1967). Its predecessor the Antarctic Treaty being the first and, in essence, both sought to prevent "a new form of colonial competition" and the possible damage self-seeking exploitation might cause (Department of State 1967). The effect of the OST was to repurpose the role of the military in outer space but it did not completely eliminate it.

The beginnings of outer space law with regard to military activity was dovetailed with civil applications and looked to custom to create its precepts. Eventually, those dictums were codified into what we now know as the OST, yet as international space law evolves its customary roots—not having been obliterated by convention—have reasserted themselves and look to take us full circle where custom not only creates new international space law but also clarifies and fills the gaps of the present jurisprudence.

This chapter will examine the development of space law with regard to military and national security activities, the effect of treaty law on the role of the military, and the evolution and shifting of how outer space is viewed in the context of the military. The discussion will conclude with an examination of the shift of how space law is coming full-circle.* This chapter is not meant as an exhaustive treatise on the subject but is intended to highlight the significant societal challenges when considering military activity in space; it also aims to provide a fundamental understanding of the relevant law and issues as well as clearing up some misconceptions about the military and national security activities in outer space.

9.2 POST-*SPUTNIK* AND PRE-OST

9.2.1 The Eisenhower Policy on Outer Space

The Cold War prompted both the United States and the Union of Soviet Socialist Republics (USSR) to look to outer space for military use. In particular the launch of *Sputnik-1* by the Soviet Union brought the standing of outer space to the forefront.† The Eisenhower administration took a hard look at the realm of outer space and noted in its preliminary national policy on outer space the significance of outer space on U.S.

* It should be noted that there is state practice that is in furtherance of international legal obligations, including treaties, and state practice in the absence of clearly defined international law that in fact can create international legal obligations, that is, custom. The distinction of when state practice is in furtherance of clear international legal obligations and when state practice creates new international law is not always apparent.

† The launch of *Sputnik* by the Soviet Union on October 4, 1957 was a game-changer in that not only did it usher in the use of outer space but also represented the beginnings of space law via customary law. It also was likely to have reinforced the perceptions Eisenhower had for use of outer space for military purposes.

national security noted "[t]he effective use of outer space by the United States and the Free World will enhance their military capability" (National Planning Board 1958, p. 6, para 17). The policy also recognized "... [o]uter space activity and scientific research would have both military and non-military applications" (National Planning Board 1958, p. 7).

The uses recognized by the policy included ballistic missiles to include Intercontinental Ballistic Missiles (ICBMs) and Intermediate Range Ballistic Missiles (IRBMs), missile defense against both ICBMs and IRBMs (National Planning Board 1958, p. 7).

The policy also incorporated applications such as military reconnaissance, weather satellites, military communications satellites, satellites for electronic counter-measures, satellites for aids for navigation, manned maintenance and resupply space vehicles, bombardment satellites, and manned lunar stations for communications relay and reconnaissance with the potential as a staging area for launching missiles to Earth (National Planning Board 1958, pp. 7–8).

Several legal issues vexed the administration about these potential uses of outer space. Prior to the launch of *Sputnik-1*, the administration was considering the use of satellites for reconnaissance of the Soviet Union, which was based on the recommendations of the so-called "Surprise Attack Panel" (later termed the Technological Capabilities Panel—TCP) on February 14, 1955.* The proposal was to establish the concept of the freedom of passage of a satellite over the territory of a sovereign nation (Terrill 1999, p. 4).

9.2.2 Freedom of Passage of a Satellite and Military Reconnaissance

Prior to the launch of *Sputnik-1*, the Soviet Union did not limit its sovereignty to the stratosphere and regarded outer space above its territory

* "The 42-member Technological Capabilities Panel of the Science Advisory Committee of the Office of Defense Mobilization was formed in response to a Presidential request made to the committee at a White House meeting on March 27, 1954, a study be made of U.S. technological capability to reduce the threat of surprise attack. The resulting panel, frequently referred to as the Killian Committee after its director, Dr. James R. Killian, Jr., interpreted its mandate broadly. The committee set as its objective an examination of the current vulnerability of the United States to surprise attack and an investigation of how science and technology could be used to reduce vulnerability by contributing to the following five developments: An increase in U.S. nuclear retaliatory power to deter or at least defeat a surprise attack; an increase in U.S. intelligence capabilities to enhance the ability to predict and give adequate warning of an intended surprise attack; a strengthening of U.S. defenses to deter or blunt a surprise attack; the achievement of a secure and reliable communications network; and an understanding of the effect of advanced technology on the manpower requirements of the armed forces." Department of State, Office of the Historian, "*Report by the Technological Capabilities Panel of the Science Advisory Committee*," footnote * on page 209, available at https://history.state.gov/historicaldocuments/frus1955-57v19/d9.

as part of its sovereign control (Terrill 1999, pp. 27–30). The launch of *Sputnik-1* challenged this claim of sovereignty as the satellite would have clearly violated the "territory" of other nations under their theory. The Soviets, when confronted with this conundrum, tried to explain *Sputnik-1* had not violated the territory of other nations as it did not pass over the territory of those nations, but rather the territories of other nations passed beneath *Sputnik-1* (Terrill 1999, pp. 27–30). The subsequent launch of *Sputnik-1* on October 4, 1957 provided the United States the opportunity to assert the principle of freedom of passage of a satellite, and the Eisenhower administration utilized customary international law to establish the legal right to not only deploy satellites for scientific purposes but for military purposes as well.* The Eisenhower administration used this dichotomy between scientific and military to cover activities under the CORONA program which was created by a small segment of personnel from the CIA, Air Force, and private industry. They were charged with finding a means to take broad imagery of the USSR to identify strategic targets, including missile launch sites and production facilities. CORONA was sanctioned by President Eisenhower in February of 1958. The public face of Corona became known as the Air Force "Discoverer" program (CIA Archives, 2015. "CORONA: Declassified").

While the issue of freedom of passage was opportunely solved for the Eisenhower administration's goal of military reconnaissance, another issue was not so readily solved. Indeed, the launch of *Sputnik-1* and the creation of the customary rule of free passage through outer space did not end the issue of reconnaissance from outer space. The Soviet Union made an unsuccessful attempt to dilute the legal effect of *Sputnik-1* by including a prohibition for reconnaissance from outer space for national security purposes in their draft submission of the Declaration of the Basic Principles

* Customary international law is defined as international obligations arising from established state practice, as opposed to obligations arising from formal written international treaties. It consists of two components. First, there must be a general and consistent practice of states. This does not mean that the practice must be universally followed; rather, it should reflect wide acceptance among the states particularly involved in the relevant activity. Second, there must be a sense of legal obligation, or *opinio juris sive necessitatis*. In other words, a practice that is generally followed but which states feel legally free to disregard does not contribute to customary law; instead, there must be a sense of legal obligation to the international community. States must follow the practice because they believe it is required by international law, not merely because they think it is a good idea, or politically useful, or otherwise desirable. The definition of customary international law is nuanced because not all states are equal when considering whether a state's practice and *opinio juris sive necessitatis* reaches the level of customary international law. See *United States v. Bellaizac-Hurtado*, 700 F.3d 1245, 1252 (11th Cir. 2012).

Governing the Activities of States in Pertaining to the Exploration and Use of Outer Space submitted to the United Nations on September 10, 1962. The Soviet Union's Draft asserted "The use of artificial satellites for the collection of intelligence information from the territory of foreign states is incompatible with the objectives of mankind in its conquest of outer space" (c.f. USSR Draft Declaration, A/AC. 105/L.2, 1962 with UN G.A. Res 1962, 1963).

9.2.3 Delineation of the Air/Space Boundary

This other issue had to do with the delimitation of the boundary of where sovereign airspace ends and outer space begins. Delimitation as anticipated under international law would take one of two forms: functional or spatial. The functional approach views outer space and airspace as a single above-ground space that does not require delimitation. Regulation of outer space and airspace under the functional approach is based on the nature of the activities versus distance from the earth. This would make a satellite that has completed one orbit of the earth and continues to orbit the earth generally traveling through what is accepted as "outer space" up and until it reenters the atmosphere. A high-altitude balloon or fixed-wing aircraft, which by their nature are limited to atmospheric flights, would be considered traversing "air space" even though they may be able to reach the lower limits of "outer space."

This approach can be expressly affected by advances in technology and engineering and was of concern to the Eisenhower administration but also favored by the Eisenhower administration in NSC 5814/1 even though it was not a publically held position (National Planning Board 1958, p. 4, para 10(b)). The spatial approach focuses on the main differences between the legal regimes of outer space and airspace and the follow-on need for definitions of (1) the spatial limitations which determine the extent of application of the principle of freedom to explore space, on the one hand, and (2) of state sovereignty over national airspace on the other (see Listner 2012). Certain laws govern travel through what is internationally recognized as "air space," but when a craft reaches a specified spatial distance that delimitates the boundary of "outer space," a different set of international regulations apply. The drawback of applying the spatial approach, aside from the regulatory and sovereignty issues, is solar activity can cause the atmosphere to expand and contract to the extent a line of demarcation may not represent a true delineation of where the atmosphere ends and outer space begins. This is reflected in, 51U.S.C. § 50902(11), which has defined a "launch vehicle" as both a vehicle built to operate in, or place a payload or human beings in, outer space; and a suborbital rocket (see Listner 2012).

As noted by NSC 5814/1, "[a]lthough the successful orbiting of earth satellites has raised a question of national sovereignty *ad coelum* and as to the doctrine of "freedom of space," the United States has not recognized any upper limit to sovereignty. In order to maintain (a) flexibility in international negotiations with respect to all uses of "space" and pending an international control agreement, (b) freedom of action with respect to the military uses of "space," the United States has taken no public position on the definition of outer space" (National Security Planning Board 1958 at p. 4, para 11). This public stance has changed little as the United States does not view the topic a practical matter to be addressed by United Nations (see Department of State 2001), but has not prevented "outer space" being defined by federal agencies and other nations.*

The Policy Statement also recognized "[t]he issue of rights in outer space was certain to be a topic at the UN General Assembly in September 1958 as well as in international dialogues after activities of the International Geophysical Year. Therefore, it would be prudent for the United States to have a prepared position on the definition of outer space, to which the Policy Statement formally recognized as being divided into two regions: "air space" and "outer space" where "outer space" is regarded as adjacent to "air space" with the lower limit of "outer space" being the upper limit of "air space" (National Security Planning Board 1958 at p. 5, para 12–13).

The National Aeronautics and Space Act of 1958 appears to mirror this position and implies the earth's atmosphere equates with the notion of "airspace" whereas conversely outer space is the area outside of the earth's atmosphere.† The legal problem of delimitation correlated to the Eisenhower

* Despite there being no international legal demarcation for outer space or interest by the United States to recognize one, the United Nations Office of Outer Space Affairs leaves the establishment of a legal boundary to individual states in their domestic law, many of which, including Australia and most recently Denmark, have applied a spatial approach of approximately 100 kilometers (~62 miles.) Moreover, the United States Air Force appears to apply a functional approach to what defines outer space by establishing outer space begins where an object can maintain orbit (around 130 kilometers or approximately 70 miles). NASA on the other hand, when awarding astronaut wings appears to apply a spatial approach recognizing a demarcation of 92.6 kilometers or (57.5 miles) see Sellers (2004) and von der Dunk (1998) for more details.

† "The National Aeronautics and Space Act was enacted on July 29, 1958 for the purpose of providing for "research into problems of flight within and *outside* the earth's atmosphere, and for other purposes." This fundamental distinction is reflected in a number of other provisions where the terms "aeronautical" respectively "space activities" figure prominently. In other words, the Space Act implies that the earth's atmosphere equates with the notion of "airspace" whereas conversely outer space is the area outside of the earth's atmosphere" (von der Dunk, 1998 at p. 257).

administration's goal to develop missile defense but also related to military reconnaissance in that while it had been established there is freedom of passage in outer space, legally delimiting outer space would be premature and not allow for the rapid development of technology and might otherwise restrict the use of future capabilities (National Security Planning Board 1958 at p. 4, para 10).

As stated above, the problem the Eisenhower administration faced was—as scientific knowledge and technology progressed—the ability to perform national activities could be implicated. The higher a nation's sovereign airspace the higher it can legally defend its airspace and perform national security activities under the purview of national sovereignty as opposed to international law. Legal delimitation also provides a potential adversary, who would be expected to develop their technology to fly higher, to exercise the legal definition of the boundary of outer space to exercise means other than the free passage of outer space to challenge the sovereignty of another nation.

In terms of ground-based missile defense, a legal delimitation of outer space would limit the ability of ground-based assets to intercept an incoming ICBM or IRBM. If the intercept is beyond the legal demarcation of a nation's sovereign airspace, the argument could be made the intercept violated the right of free passage in outer space. Article 51 of the UN Charter states in part "Nothing in the present Charter shall impair the inherent right of individual or collective self-defense if an armed attack occurs against a Member of the United Nations, until the Security Council has taken measures necessary to maintain international peace and security." It could be argued an intercept was exercising this right even though an intercept may or may not have occurred in outer space. The burden would be to show the intercept did not occur in outer space and a legally defined boundary of outer space would be not be helpful to the state defending the intercept.

9.2.4 Project Horizon: A Military Lunar Outpost

Just like the delimitation of outer space, another proposed goal of NSC 5814/1 was to establish a manned lunar station (see National Security Planning Board 1958 at p. 7, para 19(b)(4)). The prospect for a manned lunar station was manifested in Project Horizon, which was a study by the U.S. Army to establish a military lunar outpost (United States Army 1959 at p. 1, para 3). The stated purpose of Project Horizon was to demonstrate the United States' scientific leadership in outer space to support scientific explorations and investigations, extend and improve space reconnaissance and surveillance capabilities and control of space, extend and improve communications, and serve as a communications relay station. This was to be carried out to provide a basic and support research laboratory for space research and development activity, and to develop a stable low-gravity outpost for use as a launch site for deep space exploration. In addition, it was to provide an opportunity for scientific exploration and development of a space mapping and survey system, as well as provide emergency staging areas and a rescue capability or navigational aid for other space activities (United States Army 1959 at p. 2).

Critically, the study makes the point, "...from the viewpoint of national security, the primary implications of the feasibility of establishing a lunar outpost is the importance of being first. Clearly the U.S. would not be in a position to exercise an option between peaceful and military applications unless we are first. In short, the establishment of the initial lunar outpost is the first definitive step in exercising our options" (United States Army 1959 at pp. 3–4). Thus, the military's focus on a lunar outpost was not focused on the establishment of a purely armed installation and gaining territory as an end in itself but instead represented the ability to choose how the installation was utilized by establishing it before the Soviet Union created their own outpost and potentially made territorial claims. This viewpoint was potentially exacerbated by the original legal position of the Soviet Union regarding outer space prior to the launch of *Sputnik-1* (see Terrill 1999 at p. 4).

Project Horizon recognized while the technical means were available to create the outpost, there were also legal questions to be answered as space law was still in its formative stages. The Project Horizon Report itself noted "[t]here are no principles or legal rules which can be said to recognize or create any rights in or duties on the part of states operating beyond the atmosphere in outer space or on the lunar surface" (United States Army 1959 at Appendix B, p. 73). It was recognized certain precedent existed in international law, including Antarctica, whereby it was concluded "...there is neither controlling nor governing law or principle to limit or circumscribe American activity, nor basis for any other nation asserting a bar to American activity in the new dimension" (United States Army 1959 at Appendix B). In other words, the Project Horizon Report recognized while space law was still a nascent area of law, the military could not act with impunity because of the applicability of current international law. At the same time that same body of law did not prevent the establishment of the proposed lunar outpost. This goes to the point existing international law applied to the otherwise "lawless" domain of outer space, to which military and other national security activities would be subject to.

All this goes to say before the OST codified practices in outer space, many of which were created by custom, the role of the military was ambiguous from the perspective of national security. There were no clear-cut norms defining the military's role or restricting its activities aside from the dictates of national space policy; actual norms for the use of outer space and space law were being developed through custom as capabilities to utilize outer space became available. That would change with the enactment of the OST, which brought the role of the military in outer space into more focus in some respects but also created more ambiguities in terms of its activities in that domain and opened the door as to the legal nature of outer space in general.

9.3 THE OST AND THE MILITARY

While military and national security activities in general operated in a relatively vacuous legal environment after *Sputnik*, the signing and entering into force of the OST 51 years ago removed some of the ambiguity and solidified some concepts such as the freedom of access and passage to outer space. For its part, the Air Force advocated for and the United States initially adopted an *ad hoc* approach to the creation of space law via custom, which would allow practice and technology to drive the evolution of the law as opposed to academic theories (Terrill 1999 at p. 75).

Eisenhower for his part perceived outer space as a potential Pearl Harbor even before the launch of *Sputnik* and sought to meld space exploration, disarmament, and the creation of international law through his idea of "space for peace" and an environment free from national military rivalries (Terrill 1999 at pp. 3–9). This led to his proposal for a new international treaty in order to prevent a new form of colonial competition in outer space (Department of State 1967). Eisenhower's idea "space for peace" did not detract from his views of space for national security as is evidenced by both his administration's space policies in NSC 5814/1 and NSC 5918/1.

9.3.1 Peaceful Purposes and the Military

The OST would be modeled after the Antarctic Treaty of 1959 and would contain similar provisions.* In essence, the Antarctic Treaty and the OST are siblings as both contain similar principles. For example, both the OST and the Antarctic Treaty espouse the *res communis* principle† found in the preamble to the Antarctic Treaty and Article I of the OST. Additionally, both treaties state the use of their respective realms "for peaceful purposes."

Both treaties also spell out prohibitions on aggressive military activities. For example, the Article I, Paragraph 1 of the Antarctic Treaty states "There shall be prohibited, inter alia, any measures of a military nature, such as the establishment of military bases and fortifications, the carrying out of military maneuvers, as well as the testing of any type of weapons." Likewise, Article IV, Paragraph 2 of the OST stipulates "[t]he establishment of military bases, installations and fortifications, the testing of any type of weapons and the conduct of military maneuvers on celestial bodies shall be forbidden." It's noteworthy, however, both Treaties permit the use of military personnel for scientific research or for any other peaceful purposes.

It is this delineation of military activity in the OST that is most significant to military and *inter alia* national security activities. While at first glance Article IV of the OST appears to forbid military activity in its entirety it does not prohibit the use of military personal or facilities for

* The relation of the OST to Antarctica was recognized in *Beattie v. United States* where the United States Court of Appeals for the District Court for the District of Columbia and subsequently the Circuit Court of Appeals, District of Columbia had to decide whether it had subject matter jurisdiction to hear a claim for tort pursuant to the Federal Torts Claim Act that arose in Antarctica. The Court when questioning whether it had jurisdiction referred to Article VIII of the OST and recognized: "The legal status of Antarctica has been most frequently analogized to outer space. United States spokesmen suggested the 1959 Antarctic Treaty as a possible model for an OST during initial formulation discussions in 1965 and 1966. Obviously, the provisions of a treaty relating to outer space are only relevant to the present case by analogy. However, they are instructive as to the way in which the United States has acted with reference to sovereign immunity and liability for acts of its agents in a context very similar to Antarctica." *Beattie v. United States*, 756 F.2d 91, 99 (D.C. Cir. 1984).

† *Res communis* is a concept derived from Roman property law which refers to the light and the air. See Merriam-Webster Dictionary. See also, Black's Law Dictionary, Sixth Edition, *res communes*—"In the civil law, things common to all; that is, those things which are used and enjoyed by everyone, even in the single parts, but can never be exclusively acquired as a whole, for example, light and air." The idea behind *res communis* in the reference to both the Antarctic Treaty and the OST is that no sovereign can extend [state] ownership much in the same way no one can extend control over the air or the light. In other words, in the case of outer space and celestial bodies, they belong to no nation. It is notable in regard to usage and passage, that the high seas are considered *res communis*.

outer space activities so long as they are for peaceful purposes. Recalling the U.S. Army study to establish a military lunar base, Article IV of the OST would appear to forbid the establishment of the base but it could be argued Article IV would not forbid the establishment and use of the base by the military so long as it was utilized for peaceful purposes. The question is, with regard to the military activities, what constitutes a "peaceful purpose" as envisioned by Article IV. The absence of language in the first paragraph of Article IV to restrict activities in outer space to "peaceful purposes" has caused much debate over the proper interpretation of the OST, but states have generally agreed activities in space should be confined to peaceful purposes (Parkenson 1987 at p. 81).

There are two general positions on the meaning of "peaceful purpose" in Article IV: the majority position that permits military activity and the minority position that excludes military activity. The Eisenhower approach to "space for peace" policy and the directives in both NSC 5814/1 and NSC 5918/1 is one example prior to the OST of the duality of outer space. This was also exhibited when Sen. Albert Gore, representing the United States before the UN General Assembly in 1962, emphasized the point "the test of any space activities must not be whether it is military or nonmilitary, but whether or not it is consistent with the United Nations Charter and other obligations of law" (Parkenson 1987 at pp. 29–30). "In other words, the view of the United States is 'peaceful purposes' is not inconsistent with those provisions in the UN Charter and in customary international law that preserve the right of States to take armed action for their individual and collective self-defense" (see Parkenson 1987 at pp. 29–30 and also UN Charter, Articles 1, 2, 51). This means the OST does not explicitly prohibit military activity, so long it is for activities related to self-defense. What constitutes self-defense is, however, subjective and for the State to interpret.

Conversely, there is the view "peaceful purposes" means nonmilitary. "This interpretation of peaceful purposes focuses on the more general articles of the treaty and concludes that the general purpose of the treaty is to ensure that outer space is used only for peaceful purposes and for the benefit of all mankind to the exclusion of military purposes" (Parkenson 1987 at p. 83). The argument that "peaceful purposes" means nonmilitary centers its analysis on the preamble of the OST, specifically "[r]ecognizing the common interest of all mankind in the progress of the exploration and use of outer space for peaceful purposes," Articles IX, X and XI, which urge international cooperation in the peaceful exploration and use of outer space and also highlights Article I whereby "[t]he exploration and use of outer space, including the moon and other celestial bodies, shall be carried out for the benefit and in the interests of all countries, irrespective of their degree of economic or scientific development, and shall be the province of all mankind" (Parkenson 1987 at p. 83 and Article I Outer Space Treaty).

9.3.2 Article IV and Military Lunar Bases

The majority view the OST does not prohibit military activities so long as they are for peaceful purposes, including a state acting in self-defense, would have affected the aforementioned Project Horizon. As mentioned earlier Article IV would appear to have prohibited the construction of the proposed installation by the U.S. Army and its staffing with military personnel, especially if missiles or other offensive weapons were placed on it. Conversely, while the establishment of a military base on the Moon (or another celestial body) is prohibited by Article IV, when read in conjunction with Article 51 of the UN Charter, there is an exception for a military installation established for exclusively peaceful purposes and one that would not exclude the emplacement of defensive measures.

It's noteworthy that Article IV does not prohibit any offensive weapons placed in orbit of the Earth (or the Moon or another celestial body), particularly if they are designed as part of a state's defense and/or to create a deterrent effect in order to maintain the peace. (See Outer Space Treaty, Article 4, para. 2, which stipulates the prohibition of placement of weapons on celestial bodies but does not mention the orbit of a celestial body or the orbit of Earth.) Thus, a weapon system such as the "rods from god" concept* placed in orbit around the Earth or a weapon system placed in orbit around the Moon would, in all likelihood, not violate a strict reading of Article IV to the extent the placement of the system would create a deterrent effect. Likewise, the deployment of space-based missile defenses, including concepts such as Brilliant Pebbles,† would not implicate Article IV because there is no express prohibition of placement of such a system in Earth's orbit as it is intended as a defensive measure and is consistent with the right to self-defense in Article 51 of the UN Charter.

* The "rods from god" concept employs the use of a spacecraft in orbit equipped with rods formed from tungsten or some other hard substance, which would be deployed to strike a ground target. The resulting release of kinetic energy and not an explosive (conventional or nuclear) would destroy the target. See Preston et al. (2002) for more details.

† Brilliant Pebbles is a space-based interceptor concept developed as part of the SDI, which would intercept nuclear warheads before they entered their terminal phase and reentered the atmosphere. See Baucom (2004) for a discussion of Brilliant Pebbles as envisioned by the Strategic Defense Initiative.

Article IV's allowance also extends to the testing and deployment of anti-satellite (ASAT) weapons, but arguably they would create a duty under Article IX, although it is uncertain when the duty would occur. (See Outer Space Treaty, Article IX.)* The underlying fundamental principle for purposes of Article IV is the placement of weapons in orbit around a celestial body, including the Earth and the Moon, is not prohibited nor are weapons in general.

9.3.3 Nuclear Weapons in Outer Space

The prohibition against the placement of nuclear weapons found in Article IV, para 1, is one of the few explicit prohibitions in the OST. It is clear the placement of nuclear ordnance in orbit, or on or around a celestial body is prohibited, but like the other provisions of Article IV (and the OST itself) there is ambiguity to this proscription to the extent placement of nuclear weapons does not complete an entire orbit, to which the argument could be made there was no placement of a nuclear device in orbit or in outer space. This would account for the use of ballistic missiles with nuclear warheads, which traverse space at the peak of their ballistic trajectory before reentering the atmosphere without achieving a complete orbit. It also permitted the use of nuclear weapons using fractional orbital bombardment where a nuclear device is launched into space but does not make a complete orbit before reentering and striking a target (see Eisel 2005 for a technical discussion of the fractional orbital bombardment system developed by the Soviet Union. See also, Paine, 2018 for a legal and political discussion of FOBs and Article IV).

This ambiguity to the bar on nuclear weapons also brought into question the concept of the X-ray laser system envisioned as part of the strategic defense initiative (SDI). The X-ray laser system would have implicated Article IV as the system would have consisted of a small nuclear device launched aboard an ICBM or submarine-launched ballistic missile (SLBM). When detonated in outer space it would have channeled a fraction of the energy released into high-intensity laser beams, which would destroy enemy missiles during their boost phase. The device would be destroyed in the course of detonation and it is this operation of a nuclear weapon

* If a State Party to the Treaty has reason to believe that an activity or experiment planned by it or its nationals in outer space, including the Moon and other celestial bodies, would cause potentially harmful interference with activities of other States Parties in the peaceful exploration and use of outer space, including the Moon and other celestial bodies, it shall undertake appropriate international consultations before proceeding with any such activity or experiment.

in outer space—and not the resulting laser beam—Article IV technically would have been prohibited (see Parkenson 1987 at pp. 86–89).

It could be argued the devices were purely defensive and not actually be placed in orbit, but rather are of a transient nature and only remain in outer space long enough to fulfill their defensive function against incoming enemy missiles. Because the function of the X-ray laser system would have been defensive and hence nonaggressive, it could be argued its use would have correlated with Article IV's principle that outer space should be used for peaceful purposes. Conversely, the operation of a nuclear weapon even for a defensive purpose could be argued to violate Article IV considering the detonation would do more than just destroy other nuclear weapons and would in fact create collateral damage in orbit (Parkenson 1987 at pp. 86–89).

In aggregate, Article IV is not an absolute bar to military or national security activities and "permits states to use all of outer space for whatever military purposes they deem necessary, so long as it is not for an aggressive purpose (Article III), and so long as it does not involve stationing nuclear weapons or weapons of mass destruction" in outer space (Parkenson 1987 at p. 86). Simultaneously, Article IV obligations and limitations are tempered by international law in general, including the right to self-defense under Article 51 of the UN Charter.

9.3.4 Military and Article IX of the Outer Space Treaty

While Article IV bears the brunt of scrutiny when analyzing military or national security activities, one provision, Article IX, is quietly ignored. Article IX is one of the lesser-discussed provisions of the OST, especially in the context of military activities and is the least understood provision of the OST. Article IX is contained within one paragraph, but this one paragraph contains two distinct legal obligations and one legal right. The ambiguity of Article IX is preferable considering both the United States and the Soviet Union were exploring and continue to investigate the potential uses of outer space for national security activities, which might otherwise have been precluded by a concise Article IX.

A breakdown of the three distinct parts of Article IX follows. The first legal obligation of Article IX states: "States Parties to the Treaty shall pursue studies of outer space, including the moon and other celestial bodies, and conduct exploration of them so as to avoid their harmful contamination and also adverse changes in the environment of the Earth resulting from the introduction of extraterrestrial matter and, where necessary, shall adopt appropriate measures for this purpose."

The duty to prevent harmful contamination dovetails with the legal right of exploration in Article I whereby: "[t]here shall be freedom of scientific investigation in outer space, including the moon and other celestial bodies, and States shall facilitate and encourage international cooperation in such investigation." The first legal duty requires the states to perform their Article I investigations in a manner which prevents harmful contamination of extraterrestrial environments, and also to perform their activities so as to prevent the contamination of Earth from extraterrestrial matter introduced into the Earth's biosphere (see Outer Space, Article IX). This legal obligation is executed through protocols, including NASA's Planetary Protection Protocols (see NASA Policy Instruction, NASA Policy on Planetary Protection Requirements for Human Extraterrestrial Missions). It should be noted, however, there is no definition of what neither constitutes "harmful contamination" nor is there an international agreement as to what it means nor is there a customary definition. The contours of "harmful contamination" remain unclear (see Williamson 2010 for a discussion on what constitutes "harmful contamination" for purposes on Article IX).

The second legal obligation and the legal right in Article IX are cojoined and have little or no precedent in the international arena.* The second legal obligation in Article IX states: "If a State Party to the Treaty has reason to believe an activity or experiment planned by it or its nationals in outer space, including the moon and other celestial bodies, would cause potentially harmful interference with activities of other States Parties in the peaceful exploration and use of outer space, including the moon and other celestial bodies, it shall undertake appropriate international consultations before proceeding with any such activity or experiment." This segment of Article IX creates a legal duty upon a state to consult with the international community, presumably through the United Nations and specifically the Committee on the Peaceful Uses of Outer Space (COPUOS), in the event a state believes its planned space activities, including those by nongovernment actors, could potentially cause harmful interference with

* The Air Force many have inadvertently contributed to the development of Article IX with Project West Ford in 1958. West Ford studied the possibility of dispersing 75 and then 100 pounds of small metallic strips in orbit to create a space-based network, which would enable worldwide communications. The concern over the potential effect of the metal strips on the outer space environment in particular for radio and optical astronomy led to the eventual adoption of the consultation duty in the OST. See Terrill (1999 at pp. 58–63).

the peaceful exploration or use of outer space by other state actors or their nationals.

Unlike the first legal obligation of Article IX, this legal obligation has never been invoked by any state preceding a potentially damaging outer space activity. This is the case even though there have been substantial outer space activities by both the United States, the Soviet Union/Russia, and the People's Republic of China, which could have warranted its application—in particular the test of anti-satellite (ASAT) weapons. As Article IX was not invoked prior to these activities opens the question whether by not invoking the legal duty to consult for these activities a customary practice relating to Article IX has been created that reflects not when the duty to consult needs to be invoked but rather when it does not.* The idea behind this is during national security/military activities in outer space, which could potentially require the Article IX duty to consult be invoked, the consistent state practice in not invoking the legal duty to consult and the intent to be legally bound to the practice not to consult on national security activities creates a state practice of Article IX that relates to national security/military activities. The result of this is not to disregard the duty to consult but rather create state practice that excludes national security activities.

It is notable the 2007 ASAT test performed by the People's Republic of China (PRC) was claimed by Beijing to be consistent with international law, and likely so as Article IV would have permitted it. China also may have relied on the precedent created by the ASAT activities of the United States and the Soviet Union during the Cold War to justify not only its ASAT test but also its decision not to invoke the consultation requirement in Article IX.

---

* The original intent of Article IX to obligate states to consult prior to activities (presumably military activities) that could create "harmful interference" is for the most part nullified as no state wants to be the first to invoke the legal obligation and create a customary usage of not only what "harmful interference" is but when the legal duty to consult should be invoked. None of the Big Three have obligated themselves to the consultation duty in Article IX. This is not surprising because each of these states have sufficient prestige in outer space activities to establish by their own actions a customary practice of when the legal obligation to consult under Article IX is triggered. This in turn would create an international litmus test for the duty to consult under Article IX, which could expose national security activities in outer space to unwanted scrutiny. Likewise, the right to request a consultation is unlikely to be invoked by spacefaring nations for the simple fact it would invite abuse by geopolitical competitors who would question an activity merely to embarrass or harass a geopolitical competitor.

To further illustrate this effect on Article IX, the United States prior to *Operation Burnt Frost* sent a representative from NASA's Orbital Debris Office to brief COPUOS on the impending intercept attempt of USA-193. Significantly, the United States did not invoke the consultation requirement in Article IX when this consultation was made, which lends further support to a customary practice of when the legal duty to consult need not be invoked (but see Mineiro 2008, for an analysis of Article IX and its applicability to the FY-1C ASAT test and the USA-193 intercept.)

Consequently, while the military—in particular the Air Force—may have inadvertently instigated the creation of Article IX and its legal duty to consult through Project West Ford, its activities and the national security activities of other agencies and nations in orbit to include ASAT tests created an *ad hoc* rule of the legal duty through state practice that has legal recognition.

Notwithstanding the observance of the first legal duty by state space actors, the fact the second legal obligation has not been invoked gives Article IX the false reputation as a dead provision of the OST when in fact its applicability to military and national security activities has been refined by custom.

A final element in the analysis of Article IX is the legal right to request a consultation in Article IX, which dovetails the legal obligation to consult. The right is created in Article IX whereby: "[a] State Party to the Treaty which has reason to believe an activity or experiment planned by another State Party in outer space, including the moon and other celestial bodies, would cause potentially harmful interference with activities in the peaceful exploration and use of outer space, including the moon and other celestial bodies, may request consultation concerning the activity or experiment" (Outer Space Treaty at Article IX). The legal right to request a consultation must be made before a state performs a space activity. Japan's attempt to invoke the right to request a consultation after the infamous 2007 ASAT test is illustrative of how the right is not applied. The right to request a consultation in Article IX is for an activity being planned by another state party in outer space, not subsequent to the activity (see Outer Space Treaty, at Article IX).

9.4 FROM GLOBAL COMMONS TO DOMAIN OF WAR

As identified earlier, in the years leading up to the OST and those subsequent to it, outer space was being envisioned as a global commons as an offset to the cold war between the United States and the Soviet

Union, which was Eisenhower's impetus for proposing the Treaty to begin with (see Terrill 1999 at pp. 3–9). Nonetheless, the use of outer space for national security purposes remained an interest for both the United States and Soviet Union and this, inevitably, led to increased military activities. It is the need to preserve free access to outer space and to protect space assets that underlies national security efforts for outer space and not the desire to gain territory (c.f. United States Army 1959 with Bush 2006). Yet, crucially, the shift toward outer space becoming a domain of war instead of a global commons is not inconsistent with the principles laid out in the OST, or at least the interpretations created by domestic national space policy. This is found particularly in the development of anti-satellite weapons (ASATs) by both the United States and the Soviet Union during the Cold War.

9.4.1 ASATs and the Early Years of Space Activity

Both the United States and the Soviet Union started the development of ASATs in the 1950s and 1960s as ancillary capabilities to missile defense systems. In the case of the United States, it developed its capability and eventually deployed an operational ASAT* in response to Premier Nikita Khrushchev's threat to deploy orbital thermonuclear weapons, that is, fractional orbital bombardment on August 9, 1961 in which he stated "You [the Americans] do not have 50- or 100-megaton bombs; we have bombs more powerful than 100 megatons. We placed [cosmonauts] in space, and we can replace them with other loads that can be directed to any place on Earth" (Eisel 2005).

While Khrushchev's threat of deploying thermonuclear weapons diminished, and hence the initial U.S. ASAT response with it, the Soviet Union did deploy such a system but not until decades later (see Eisel 2005 for a technical discussion of the fractional orbital bombardment system developed by the Soviet Union). Both the Soviet Union and the United States developed dedicated ASAT capabilities in the 1950s and 1960s based on nuclear ordnance to make up for the lack of precision guidance available (Grego 2012). The United States system eventually took the form of Project

* The first operational ASAT was a variation of the U.S. Army's Nike-Zeus anti-ballistic missile (ABM), which used a 400 kiloton, W-50 thermonuclear warhead called Nike-Zeus DM-15B code-named "Mudflap." The system became operational on June 27, 1963. Mudflap was deactivated in May 1966 in favor of an ASAT based on the U.S. Air Force Thor missile. See generally, "Program 505, available at http://www.astronautix.com/p/program505.html (for a description and timeline of Program 505).

437, which was authorized by the Kennedy administration on August 6, 1963 (see Bundy 1963). At this juncture, the ASAT efforts of both the United States and the Soviet Union were directed at missile defense (the nuclear nature of the efforts notwithstanding).

The archetype of space as global commons, as envisioned by the OST, gradually moved more into the direction of a theater of war in the 1970s when the United States interest in ASATs shifted during the Ford administration from defending against fractional orbital bombardment to recognizing the military utility of outer space not only for its own interests but for those of its geopolitical and ideological adversary. National Security Directive Memorandum 345 recognized the utility of Soviet military space assets, which could be used to support Soviet forces in the event of war, and directed the development of an ASAT capability to eliminate those assets (see Scowcroft 1977).

The policy directive sought both a hard and soft-kill ASAT capability, which was classified to prevent a reciprocal Soviet response.* The Ford policy directive also instructed the Department of State, concurrent with the initiative to acquire an ASAT capability, to identify and assess any arms control measures that might restrict the development of ASATs, to raise the level of the threat environment to necessitate the use of ASATs, and spell out what actions would constitute interference with the space systems of a foreign state (see Scowcroft 1977).

9.4.2 ASATs in the 1970s

The Carter administration built on this foundation of recognizing outer space as a domain of war through its continued support of the development of ASATs but also took an active track with regard to the issue of arms control related to ASATs. The Carter administration sought to actively address the issue of ASATs with the Soviet Union as opposed to Ford administration's direct goal of developing an ASAT to use in the event

* The high priority concern for the Ford administration in Program 437, which eventually evolved into the ASM-135 or the "flying tomato can" as it was called, was Soviet electronic ocean surveillance satellites (EORSATs) and nuclear-powered radar ocean surveillance satellites (RORSATs), which surveilled United States naval vessels and in turn transmitted that information to Soviet surface ships and submarines. The point of Program 437 was not to use the ASAT capability as a bargaining chip for arms negotiation but to develop an active weapon system to be used. This coupled with the resumption of the Soviets testing their coorbital ASAT, *Istrebitel Sputnik*, convinced the Ford administration to move forward. See generally, Dwayne Day, "Smashing RORSATs: The origin of the F-15 ASAT program," *The Space Review*, January 10, 2010, available at http://www.thespacereview.com/article/1540/1.

hostilities broke out between the United States and the Soviet Union (see Carter PD/NSC-39; PD/NSC-45).

A Memorandum from Zbigniew Brzezinski, then assistant for National Security affairs dated September 23, 1977 notes the administration's desire to pursue a comprehensive treaty with the Soviet Union to ban ASATs, except those using electronic means. The Memorandum directed the continued research and development of hard-kill, kinetic ASATs but prevented operational and space-based testing (see Office of the Historian, Department of State 1977). The restriction for operational and space-based testing was short-lived and was lifted by President Carter on March 10, 1978 in Presidential Directive/NSC-33 (see Carter, PD/NSC-33).

The Carter administration's two-prong approach of seeking a verifiable ban and continuing development of an ASAT capability via Program 437 is stipulated in the Carter administration's National Space Policy (Presidential Directive/NSC-37) (see Carter, PD/NSC-37 at p. 4, para 2(d)). Additional to the continuing support to develop an ASAT, Presidential Directive/NSC-37, which is the first comprehensive national space policy since NSC 5814/1 during the Eisenhower administration, includes three salient principles, which point to the United States' growing recognition of outer space as a domain of war: PD/NSC-37 stipulates the "[c]ommitment to the principles of the exploration and use of outer space by all nations for peaceful purposes and the benefit of all mankind. 'Peaceful purposes' allow for military and intelligence-related activities in pursuit of national security and other goals" (Carter, PD/NSC-37 at p. 1, para 1(a)). PD/NSC-37 also asserts the principle "[t]he space systems of any nation are national property and have the right of passage through and operations in space without interference. Purposeful interference with operational space systems shall be viewed as an infringement upon sovereign rights" (Carter, PD/NSC-37 at p. 2, para 1(d)). Finally, PD/NSC-37 stipulates "[t]he United States will pursue activities in space in support of its right to self-defense" (Carter, PD/NSC-37 at p. 2, para 1(e)).

The fundamental reason to illustrate the policy decisions of these two administrations toward ASATs is both the Ford and Carter administration recognized the strategic value the growing use of satellites by the Soviet Union presented and the advantage remote sensing and communications would provide the Soviet Union in the time leading up to and during a conflict. Program 437, unlike its missile defense predecessor Program 505, was designed to eliminate that advantage despite the *res communis* nature of outer space and the freedom of passage established both through *Sputnik* and the OST. The Ford administration and the Carter administration took advantage of the allowance in Article IV of the OST to further national security interests up and to the point of interfering with the sovereign right of the Soviet Union to place and operate space assets in orbit.

Significantly, the Soviet Union appeared to adopt a similar policy position with regard to the OST. The development and deployment of the *Istrebitel Sputnik* co-orbital ASAT was designed to deny the United States the use of outer space for its national security assets (on a contemporary note, the People's Republic of China appears to have learned the lessons of the United States and the Soviet Union during the cold war and is vigorously pursuing its own ASAT capability). While not expressly admitting outer space had become a domain of war, the policies of both the United States and the Soviet Union tacitly acknowledged it all the while walking a fine-line to harmonize with the spirit of OST while taking advantage of the letter of the law to further national security interests.*

9.4.3 High Profile Missile Defense in the Reagan Years

The identification of outer space as a domain of war, and its harmonization with Article IV of the OST, continued and evolved in the space policies of subsequent presidents. The Reagan administration—in particular its National Space Policy put into effect on July 4, 1982—continued the trend toward recognizing space as a domain of war during the closing years of Program 437 and afterward. The Reagan administration's National Space

* The Reagan administration inherited Program 437 but took a different policy direction with regard to ASATs. The Reagan 1982 National Space Policy looked to an ASAT capability as a deterrent to the Soviet Union's *Istrebitel Sputnik* ASAT, which at that time was an operational system. (See Reagan, NSDD 42, 1982) "The United States will develop and deploy an ASAT capability to achieve an operational system at the earliest practical date. The primary military purposes of a United States ASAT capability are to deter threats to space systems of the United States and its Allies and, within such limits imposed by international law… (remainder redacted))." National Security Decision Directive Number 42 also continues and modifies the Carter policy to consider disarmament agreements by not directly referencing ASAT but instead coining the term "space weapons." This approach does not specifically mention ASATs but it does not specifically exclude them either. This posture continues the two-prong approach of the Carter administration to seek disarmament while at the same time positioning itself to deploy so-called "space weapons" should the need arise (see NSDD 42, para. I(I)). Program 437 (ASM-135) was eventually tested against the United States' satellite P78-1 (Solwind) but Congress cut funding and restricted further development not long after the test. This left the Reagan administration in a quandary as to the deployment of an ASAT and its policy position of deterrence, which led it to seek alternatives to the ASM-135. (Reagan NSSD 4-86). The study ordered by NSSD 4-86 found no viable options to the AGM-135 and the administration sought to have the Congressional restrictions lifted (see Reagan, Ronald, NSSD 258, "*Anti-Satellite (ASAT) Program*," February 6, 1987). Ultimately, Congress did not lift the testing/funding restrictions, which led the Reagan administration to cancel Program 437 in 1988.

Policy, NSSD-42, asserts as its first stated goal to "…strengthen the security of the United States…" (Reagan NSSD-42 at p. 1, para 1).

NSDD-42 also differentiates the space activities of the United States into three sectors and for the first time delineates military activities as a separate sector stating "The United States space program will comprise three separate, distinct, and strongly interacting sectors—Military, National Foreign Intelligence, and Civil. Close coordination, cooperation, and information exchange will be maintained among these sectors to avoid unnecessary duplication. All programs in these sectors will operate under conditions that protect sensitive technology and data and promote acceptance and legitimacy or United States' space activities" (Reagan NSDD-42, p. 2, para I(f)).

The differentiation of military as a separate sector complements another critical goal of NSDD-42 "[t]he United States will pursue activities in space in support of its right of self-defense" (Reagan NSDD-42 at p.3, para I(H)). Indeed, the strengthening of the security of the United States, differentiating military space activities and the stated goal of self-defense set the policy foundation for the Reagan administration's signature of the SDI research program (see Reagan NSDD-172 at p. 1). There has been some debate as to whether SDI was contrary to the ABM Treaty of 1972 by the strict reading of Article V of that accord; however, the Reagan administration posited SDI was deemed a research program intended to abide by limitations in the ABM Treaty (for further details see Harbour 1989).

NSSD-42's posture of separating out military activities signalled a willingness to interpret Article IV of the OST and the role of the military in general beyond the scope of a global commons. In doing so, the Reagan administration placed a primacy on the military's role and obligation of the protection of the United States for not only self-defense but its role in outer space in the event hostilities should break out.

9.4.4 National Space Policy and the Military Dimension in the 1990s

The national space policy of President George H.W. Bush continued to recognize the military role in outer space and reemphasized the right of self-defense and did not completely back away from the implication of outer space as a domain of war. National Security Directive 30/National Space Policy Directive 1 stipulates as one of its goals "[t]he United States will pursue military and intelligence-related activities in outer space in support of its inherent right to self-defense and its commitments to its allies" (Bush NSD-30/NSPD-1 at p. 3).

It also had a specific military policy, which reads in part "[t]he military will conduct those activities in space that are necessary to the national defense. Space activities will contribute to national security objectives by (1) deterring, or if necessary, defending against enemy attack; (2) assuring that the forces of hostile nations cannot prevent our own use of space; (3) negating, if necessary, hostile space systems; and (4) negating if necessary, hostile space systems. Consistent with treaty obligations, the military space program shall support such as command and control, communications, navigation, environmental monitoring, warning, tactical intelligence, targeting, ocean and battlefield surveillance and force application (including research and development programs, which support these functions). In addition, military space programs shall contribute to the satisfaction of national intelligence requirements" (Bush, NSD-30/NSPD-1) at p. 5, para V).

In other words, the effect of paragraph V of NSD-30 (NSPD-1) was not only to posture the role of the military for the administration, but also elucidate the administration's interpretation of Article IV of the OST to include the use of force and implicitly give credence to outer space as a domain of war.

The national space policy of President Clinton, PDD/NSC 49 (PDD/NSTC 8), appears to have taken a parallel course with respect to an overt military space policy. While emphasizing self-defense and the national security role of outer space in support of other theaters of war the policy also recognized the potential for offensive military actions even while not explicitly mentioning the role of the military (see National Science and Technology Council PDD/NSC 49 (PDD/NSTC 8) at pp. 1–2, para 3). PDD/NSC 49 also outlines "Defense Sector Guidelines" in paragraph 6, which are similar in scope to paragraph V of NSD-30/NSPD-1 (see National Science and Technology Council PDD/NSC 49 (PDD/NSTC 8) at pp. 5–6, para 6). In doing so, PDD/NSC 49 (PDD/NSTC 8) appears to have not backed off from the stance taken by the prior administration in NDS-30/NSPD-1 but rather rephrases it.

9.4.5 The Bush Space Policy of 2006

In 2006, the administration of George W. Bush provided a significant turning point for recognition that outer space is a domain of war in its National Space Policy.* Up to this point, the underlying theme "self-defense" contained an unspoken predicate there would have to be an initial hostility or a threat of hostile actions for self-defense to take place. Self-defense and

* References to the 2006 Bush National Space Policy alludes to the unclassified version released on August 11, 2006. Not even the specific number of the executive order for the 2006 National Space Policy is known at the writing of this chapter.

protection of the United States is an obligation of the military, and the military even while acting in the role of defense, does so in a theater of war. Yet, the policies of prior administrations avoided an explicit statement that outer space is a domain of war. Pragmatically, this was to accommodate the geopolitical environment at the time and potentially the desire not to create an interpretation of Article IV of the OST for the Soviet Union to use for overt propaganda purposes while covertly furthering their own military outer space activities.*

The 2006 National Space Policy was considered highly provocative for two specific reasons. First, it strayed from the path of prior administrations and excluded the possibility of negotiating or entering any legally binding arms control agreement or measures related to outer space by stating "the United States will oppose the development of new legal regimes or other restrictions which seek to prohibit or limit the United States' access to or use of space. Proposed arms control agreements or restrictions must not impair the rights of the United States to conduct research, development, testing, operations, or other activities in space for US national interests…" (Bush, U.S. National Space Policy, Unclassified, 2006, p. 2).

The second area of controversy centers on the assertive nature of the policy where "[t]the United States considers space capabilities—including ground and supporting links—vital to its national interests. Consistent with this policy, the United States will: preserve its rights, capabilities, and freedom of action in space; dissuade or deter others from impeding those rights or developing capabilities intended to do so; take those actions necessary to protect its space capabilities; respond to interference; *and deny if necessary, the use of space capabilities hostile to US national interests* {*Emphasis added*}…" (Bush, National Space Policy, Unclassified, 2006, pp. 1–2).

The 2006 National Space Policy broached the position of self-defense and proffered its stance to include active denial and as a result postured the

* After the Reagan administration announced the Strategic Initiative, the Soviet Union progressed its space weapons program in the form of the Skif-DM (*Polyus*) combat station. The station was intended to be armed with a laser and other "defensive weapons" to protect it from ASATs, but those weapons could have been utilized for offensive operations. Most significantly, some diagrams of the *Polyus* revealed it was designed to carry nuclear "space mines." While the combat station itself would not have been banned by Article IV of the OST, nuclear ordnance would have been a flagrant violation. The *Polyus* failed to reach orbit, but it had it done so it could have set off a "space weapon" race. See generally, Day, Dwayne A., Kennedy III, Robert G., "Soviet Star Wars, "The launch that saved the world from orbiting laser battle stations." *Air and Space Magazine*, January 2010 (for a history of the *Polyus* combat satellite).

United States in a more assertive attitude with regard to its rights and use of outer space. This suggests the 2006 National Space Policy was considering preemption as part of its self-defense strategy (see generally Reisman and Armstrong 2006 for a further discussion on the concept of preemption as it relates to self-defense and UN Article 51).

The response from the international community was not positive with accusations from geopolitical allies and foes alike leveling accusations the United States was seeking to dominate outer space. The policy also drew the ire of political commentators and scholars. Some stated that it provided evidence that the United States was seeking to gain a monopoly in outer space to the exclusion of others. One Chinese commentator suggested the risk arising from such dominance by one country could lead other nations—specifically China—to develop anti-satellite and space weapons to defend against that goal (see generally Shixiu 2007). Congress inquired whether the 2006 National Space Policy could have been the impetus for the PRC's ASAT test. The State Department issued a report in response on April 23, 2007 and concluded "[e]ven before issuance of the U.S. space policy, China conducted three previous tests of this direct-ascent ASAT weapon and, by September 2006, China had used a ground-based laser to illuminate a U.S. satellite in several tests of a system to "blind satellites." The report further concluded "[b]efore and after this latest ASAT test, PRC military and civilian analysts have voiced concerns about China's perceived vulnerability against U.S. dominance in military and space power. After the test, a senior colonel of PLA's Academy of Military Sciences said "outer space is going to be weaponized in our lifetime" and "if there is a space superpower, it's not going to be alone, and China is not going to be the only one" (Kan 2007).

Notwithstanding the assertive stance and the tumult created by the 2006 National Space Policy, the Bush administration stopped short of explicitly asserting space as domain of war. Yet, the implicit effect of the 2006 National Space Policy was to recognize outer space not only as a global commons, as envisioned by the OST, but simultaneously a domain of war with all of the attendant risks brought on by such an acknowledgment.

9.4.6 Contemporary U.S. Policy: Engagement and Expansion

The Obama administration effectively took a step back from viewing space as a domain of war and more toward a global commons. The administration's National Space Policy, PPD-4, was effective from June 29, 2010 and took a similar approach to self-defense and right of free passage as prior policies but stepped back from the more assertive tone of the 2006

National Space Policy. (See generally, Obama, National Space Policy of the United States, 2010.) PPD-4 also reinstituted the willingness of the United States to enter into verifiable arms control agreements but also sought to address outer space security through transparency and confidence-building measures (TCBMs)* (see Obama, National Space Policy of the United States, 2010 at p. 7).

PPD-4, like its predecessors, also sought to deter potential adversaries, but instead of active deterrence through the use of force, PPD-4, as elaborated by the 2011 National Security Space Strategy (NSS) and the questionable "passive deterrence" through the "layered approach" envisioned by the Eisenhower Institute (see generally, Stone 2015 for a discussion of the layer approach to deterrence). It is notable in the years following the issuance of PPD-4 the Obama administration likely altered its policy toward outer space security given activities of the PRC and the Russian Federation with respect to continuing development and testing of ASATs through disguised missile-defense tests and other means (see generally, Weeden 2014).

Throughout the policies surrounding ASATs and subsequent national space policies in the United States, each administration did an elaborate dance with the OST and the issue of outer space security to the point of tacitly acknowledging outer space was a global commons but also existed as a domain of war. In essence, "space for peace" envisioned by Eisenhower continued to play out in the refusal by successive administrations to expressly acknowledge space is also a domain of war and doubtless to the frustration of warfighters much in the same way the military felt its obligation to defend the United States was constrained by Eisenhower's views (see Terrill 1999 at pp. 43-45, 48).

* Transparency and confidence-building measures (TCBMs) are part of the legal and institutional framework supporting military threat reductions and confidence building among nations. They have been recognized by the United Nations as mechanisms offering transparency, assurances, and mutual understanding among states and they are intended to reduce misunderstandings and tensions. They also promote a favorable climate for effective and mutually acceptable paths to arms reductions and nonproliferation. The General Assembly at its 73rd plenary meeting on December 7, 1988 endorsed the guidelines for TCBMs decided upon by the Commission on Disarmament on December 12, 1984. TCBMs have been used extensively for purposes of arms control and specifically in the arena of nuclear weapons. However, when applied to space activities TCBMs can address other space activities outside of those performed for by the military or for those performed for national security reasons. See, Andrey Makarov, *Transparency and Confidence-Building Measures: Their Place and Role in Space Security,* Security in Space: The Next Generation-Conference Report, March 31–April 2008, United Nations Institute for Disarmament Research (UNIDIR), 2008. See also, UN General Assembly, 43rd Session, 1988, *Guidelines for confidence-building measures* (A/43/78H).

9.4.7 Space as a Domain of War?

The first explicit recognition of outer space as a domain of war was broached not by the United States but by the PRC. China's 2015 Defense White Paper took the bold step of asserting outer space as a critical security domain along with the oceans, cyberspace, and its nuclear force (China's Military Strategy 2015 at p. 4). A 2017 report to Congress notes this assertion is more than mere doctrine. The 2017 report remarks that China's space and counter-space operations fall under its Strategic Support Force, which was established by the People's Liberation Army (PLA) in 2015 to direct (the PLA's) space, cyber, and electronic warfare missions (Office of the Secretary of Defense, Military and Security Developments Involving the People's Republic of China 2017, p. 34). While the report notes there is little information publically available about the Strategic Support Force, it can be construed its mission—as related to outer space—harmonizes with the PLA's view of outer space as a "commanding height" (see China's Military Strategy 2015 at p. 4). This suggests that the PLA views outer space as a domain of war in support of its terrestrial military objectives and not the focal point of war itself.

The Trump administration appears to be following suit as the military, in particular the United States Air Force, has evolved its view of the perception of outer space. The Air Force Secretary, Heather Wilson, remarked in written testimony before the Strategic Forces subcommittee of the Senate Armed Services Committee (SASC) on May 16, 2017 "…that space no longer is just an enabler and force enhancer for U.S. military operations, it is a warfighting domain just like air, land, and sea" (Wilson et al. 2017 at p. 2). This is a substantial shift in the way the Air Force views outer space and may portend a substantial shift in U.S. national space policy from PPD-4 and from the 2011 National Security Space Strategy, which viewed space as a "contested environment" but stopped short of calling space a domain of war (for further details see Gates and Clapper 2011 at p. 3). Secretary Wilson's statement signifies a pragmatic approach to outer space security by the Trump administration and runs in line with the view of outer space taken by China in its 2015 Defense Whitepaper.

Secretary Wilson's public comments on outer space indeed heralded a shift in policy toward outer space beginning with the Trump Administration's National Security Strategy on December 19, 2017 (see Trump, 2017, p. 31). It is notable that the strategy takes the tone of selfdefense under Article V of the UN Charter, which does not quite broach the idea of space as a domain of war. It would not be until one month later with the release of the National Defense Strategy on January 19, 2018 where the Trump Administration articulates with certainty what other administrations have refused to utter for decades: outer space is a domain of war (see Mattis, 2018, p. 6).* How this recognition will be reconciled with PPD-4 is unclear as neither the National Security Strategy nor the National Defense Strategy supplants the current National Space Policy, and the Trump Administration has no immediate plans to replace the Obama-era National Space Policy.†

The shift in doctrine that recognizes space as a domain of war suggests the role of the military and its activities in outer space will become more prominent and clarified. Indeed, beyond the domestic policy considerations, the scope of Articles III and IV of the Outer Space Treaty will be influenced as will the military's role in outer space activities by recognizing outer space as a domain of war. To the extent of recognizing outer space as a domain of war is seen as inconsistent with the Outer Space Treaty in general and Article IV specifically state practice will add another layer to or perform an end-run-around of the Outer Space Treaty to produce the desired international legal environment to allow the military to perform its obligations.

9.5 CUSTOMARY LAW AND THE FUTURE ROLE OF THE MILITARY AND SPACE LAW

The recognition of outer space as a domain of war is not an evolution of the early years of the twenty-first century. Instead it is merely a return to the beginning of the space age when the United States military realized the potential for outer space. Geopolitical events and in particular the Cold War led Eisenhower to propose the OST instead of allowing the laws

* *Space and cyberspace as warfighting domains.* The Department will prioritize investments in resilience, reconstitution, and operations to assure our space capabilities. We will also invest in cyber defense, resilience, and the continued integration of cyber capabilities into the full spectrum of military operations.

† Dr. Scott Pace, Executive Secretary of the National Space Council, gave a keynote address at the 13th Annual Eileen Galloway Symposium on Critical Issues in Space Law on December 13, 2007 in Washington, DC. In response to a question after his address, Dr. Pace remarked a new comprehensive National Space Policy to supplant PPD-4 is not a priority for the Trump Administration, but the Administration recognized the need to make an interim change in PPD-4 that would permit NASA to refocus its efforts and allow Congress the time to allocate appropriations in the FY 2019 budget for those changes. The Administration made this amendment to PPD-4 via its Executive Order, Space Policy Directive 1 on December 11, 2017. See Space Policy Directive 1, 83 Fed. Reg. 59501-59502 (December. 11, 2017).

and norms of outer space to continue to form via customary international law. Customary international law has not been silent and has affected how the gaps within the provisions of the OST have been filled for example with Article IV and Article IX.* In a sense, custom has affected these two provisions of the OST to shape them to what is desired to meet geopolitical goals.

It can be argued the OST is an accord out of its time. Critics will point to the fact it does not have the flexibility to address the rapid development of technology and the resultant capabilities for outer space activities as well as the changing geopolitical environment (Hickman 2007). This is likely one of the reasons why the U.S. Air Force preferred the law of space to develop through state practice as technology developed and the resultant customary international law (Terrill 1999 at pp. xi, 75). Even though the OST has served its role, it has become a geopolitical icon, which is politically untenable to retire or amend. Custom still provides the means to transfigure the meaning of the words or silence of the OST even if the words or the silence of the OST do not change.†

9.5.1 PAROS & PWTT: The Difficulties of New Treaties

Certainly, even before the OST was signed, proposals to address outer space security by restricting use of outer space through legally binding measures were proposed and rejected. For example, during negotiations for the Outer Space Treaty, the Kennedy Administration considered either a separate arms control measure for placing weapons in outer space or a ban on weapons of mass destruction, including nuclear weapons (see generally Kaysen, NSAM-192, "Separate Arms Control Measures for Outer Space") After the OST was signed and entered into force, discussions surfaced for drafts of proposed treaties to prevent an arms race in outer space, including an initiative in the Conference of Disarmament called the Prevention of

* More recently, the issue of Article I and II of the OST as it applies to property rights and mining of resources from celestial bodies has been the focus of efforts to use custom to create a legal posture, which creates a work-around the OST to permit the ownership of "space resources." In particular, the United States and Luxembourg have passed domestic laws that permit its citizens or companies registered in its jurisdiction to extract resources from celestial bodies, which depend on an end-run-around of Article II of the OST. See generally, 51 U.S.C. §§ 51301–51303 and "*Projet de loi, sur l'exploration et l'utilisation des ressources de l'espace.*"

† There is an argument some or most of the provisions of the OST have become custom over the 51 years since it entered into force. Thus, if a state were to withdraw from the OST, it could be argued this state would still be bound by its principles and duties.

an Arms Race in Outer Space (PAROS).* Also, within the Conference is the Treaty on the Prevention of the Placement of Weapons in Outer Space and of the Threat or Use of Force against Outer Space Objects (PPWT), which is a draft treaty by the Russian Federation and the PRC presented to the Conference of Disarmament in 2008 (as well as a redraft in 2014), which would ban space weapons. The fundamental problem with banning "space weapons" as asserted by PAROS and the PPWT is there is the lack of legal definition of what is a "space weapon" and the ability to verify whether a space object meets a legal definition because instrumentalities in outer space can perform either a benign or harmful function. This "dual use" conundrum is one of the issues of deploying space debris remediation technology, in that technology which can remove space debris can also perform as an ASAT.

More jaded observers of international relations could view the "space weapons" debate as an attempt to create a soft-power straw man. Both the Russian Federation and China are likely to be aware the PPWT will probably not be formally signed much less ratified by the United States for the simple reason verification is impracticable and likely see the PPWT as a tool to curry soft-power in the UN (see Listner and Pillai 2014). Aside from the difficulties (including verification), new, legally binding treaties for space weapons for outer space are falling out of favor with states. While space-faring nations like the United States and its geopolitical allies acknowledge and accept the OST and its children, albeit with interpretations favorable to their outer space activities, there is no appetite to either amend or add to the existing legal regime of outer space with new legally binding accords.

---

* "The Prevention of an Arms Race in Outer Space (PAROS) is a UN resolution which reaffirms the fundamental principles of the 1967 OST and advocates for a ban on the weaponization of space. The PAROS resolution acknowledges the limitations of existing laws related to outer space and recognizes the OST "by itself does not guarantee the prevention of an arms race in outer space." The resolution advocates for further measures to prevent an arms race in outer space by, among other things, urging all state parties, particularly those with space capabilities, to adhere to the objectives of PAROS. It calls on the Conference on Disarmament (CD)—the UN disarmament negotiating forum—to establish an ad hoc committee regarding PAROS resolution issues." Federation of American Scientists, "*Prevention of An Arms Race in Outer Space*," available at https://fas.org/programs/ssp/nukes/ArmsControl_NEW/nonproliferation/NFZ/NP-NFZ-PAROS.html. The Soviet Union and its successor the Russian Federation pushed to have PAROS as an item on the UN General Assembly's agenda in 1981 and its heir successfully pushed for Resolution 70/27, "No first placement of weapons in outer space" on December 7, 2015 despite the objection of the United States and Israel. See *No first placement of weapons in outer space*, G.A. Res. 27, 79th Sess., Agenda Item 95(b), UN Doc. A/RES/70/27 (2015).

With an inclination to new legally binding measures fading, the European Union proposed its Code of Conduct for Outer Space Activities in 2008. The Code of Conduct was an attempt to create nonbinding or soft-law measure addressing activities in outer space as opposed to legally binding measures, which would ban instrumentalities or prohibit certain activities in outer space that might be hostile. The original effort was unsuccessful under the direction of the EU as it did not gather enough support in the international community and was reorganized under the guidance of UNIDIR as the International Code of Conduct for Outer Space Activities (see generally Beard 2017 for an overview of the Code of Conduct and soft-law measures in general). The effort failed again in 2015 when two procedural moves doomed the effort before discussions over the 2014 draft could progress (see Listner 2015).

With most avenues for creating space law limited and certain avenues shut, the likely place for the military to look for the future of space law and policy is the past when the United States and the Air Force looked to customary international law before deviating to the OST (see Terrill 1999 at pp. xi, 75). Indeed, the Air Force, while formally acknowledging outer space is a domain of war, which is supported by national policy, also acknowledges the key to creating future international rules of conduct lies in customary international law.* However, efforts by the United States to use custom to further the law of space will not be on the same empty playing field (as it was post-*Sputnik*) and will have to contend with multilateral efforts using custom to levy norms of behavior (see Listner 2014 for details of how the specter of custom impacts on the aforementioned International Code of Conduct).

One such future customary effort that will need to be addressed are the proposed Guidelines for the Long-Term Sustainability of Outer Space Activities. These are being created by a working group of the Scientific and Technical Subcommittee of the Committee of the Peaceful Use of Outer Space (COPUOS) (see COPUOS, A/AC.105/C.1/L.308, 2017).

Two of the Guidelines under consideration and discussion will have significant consequences for national security space activities should they

* Secretary Heather Wilson commented the United States and its international partners must establish rules for the use of outer space. Secretary Wilson acknowledged "...space law will include customary laws and negotiations with others on what is acceptable." Mancione, Scott, "*Air Force has long to-do list for space operations, Wilson says*," Federal News Radio, October 5, 2017, available at https://federalnewsradio.com/air-force/2017/10/air-force-has-long-to-do-list-for-space-operations-wilson-says/.

be adopted by the United Nations General Assembly and accepted by the United States. Guidelines 7 and 8 as proposed by the Working group could alter the international framework for outer space law and form the basis of customary international norms, which supplant current interpretations of the OST, including Article IV (see A/AC.105/C.1/L.308 2017, pp. 19-20 for the text of Guidelines 7 and 8).

This could effectively pre-empt any efforts by the United States to create customary international norms and box-in the outer space activities of the military to the detriment of national security and the benefit of geopolitical competitors, which foreshadows the Guidelines, if asserted as customary international law, and would create the very legal impediment sought to be avoided by entering into legally binding treaties.

This illustrates the original conundrum the military faced with regard to creating international law: the military wanted to continue its *ad hoc* approach to creating outer space law instead of academia being the prime influencer (Terrill 1999 pp. xi, 75). Yet, after 60 years many within the space law community, including academics and practitioners are firmly entrenched in the creation of space law and seek to apply customary international law to promote their views, many of which are not favorable to military activity.*

9.6 CONCLUSION

Sixty years ago, *Sputnik-1* took the ideas of space law from the realm of conjecture and started the process of making theories reality. Part of the reality is the role the military plays in outer space activities and its obligations to national defense. The evolution of space law from custom to treaty and back to custom again signifies not only the importance of the role in the military but also its consistency with ideologies which sought and continue to seek to ban its activities altogether.

Despite the maneuvering of geopolitical adversaries to diminish the advantage of the high ground of outer space for their rivals, the course of space law prospectively will reaffirm the place and contribution the military and its activities will have on the legal regime of outer

* The Manual on International Law Applicable to Military Use of Outer Space is (MILAMOS) is a project launched by the McGill Centre for Research in Air and Space Law (CRASL), the University of Adelaide's Research Unit on Military Law and Ethics (RUMLAE), lawyers, and members of nongovernmental organizations in May 2016. It purports to seek to create a "rule-based" global order for the use of outer space by the military. The net effect of the effort is to create customary international law that could conceivably limit the use of outer space by the military, which is the very scenario the Air Force originally sought to avoid post-*Sputnik*.

space. Indeed, as the view shifts to recognizing space as a domain of war, the principles and rules of conduct consistent with that viewpoint will need to be identified to not necessarily increase the chances of military engagements in outer space but rather to provide a means for future engagements to be deterred or otherwise avoided through misunderstanding. All in all, if the wisdom from that Chinese general from the Kingdom of Wu could be conferred in person today, he would not only recognize the importance of outer space as the high ground but also its value as a domain for war and encourage and not dissuade its use by the military and for national security in general.

REFERENCES

Baucom, D.R., 2004. The rise and fall of brilliant pebbles. *The Journal of Social, Political and Economic Studies*. 29 (2), Summer 2004.

Beard, J.M., 2017. Soft law's failure on the horizon: The international code of conduct for outer space activities. *University of Pennsylvania Journal of International Economic Law*. 38, 2.

Bundy, M., 1963. NSAM-258. Assignment of Highest National Priority to Program 437. John F. Kennedy Library and Museum, August 6, 1963.

Bush, G.H.W., NDS-3/NSPD-1, 1989. *National Space Policy*. November 2, 1989. Available at https://bush41library.tamu.edu/files/nsd/nsd30.pdf.

Bush, G.W., 2006. *U.S. National Space Policy, Unclassified*. August 11, 2006. Available at http://marshall.wpengine.com/wp-content/uploads/2013/09/U.S.-National-Space-Policy-31-Aug-2006.pdf.

Carter, J., PD/NSC-33, *Arms Control for Anti-satellite (ASAT) Systems*. Jimmy Carter Library and Museum. March 10, 1978. Available at https://www.jimmycarterlibrary.gov/assets/documents/directives/pd37.pdf.

Carter, J., Presidential Directive/NSC-39, 1978. *Instructions to the U.S. Delegation to the ASAT talks with the Soviets Commencing on June 8 in Helsinki*. Jimmy Carter Library and Museum. June 6, 1978. Available at https://www.jimmycarterlibrary.gov/assets/documents/directives/pd39.pdf.

Carter, J., PD/NSC-37, "*National Space Policy*" Jimmy Carter Library and Museum, para. 1(a), May 22, 1978. Available at https://www.jimmycarterlibrary.gov/assets/documents/directives/pd37.pdf.

Carter, J., PD/NSC-45, 1979. *Instructions to the U.S. Delegation to the ASAT talks with the Soviets Commencing on January 23 in Bern*. Jimmy Carter Library and Museum, January 22, 1979. Available at https://www.jimmycarterlibrary.gov/assets/documents/directives/pd45.pdf.

China's Military Strategy, State Council Information Office of the People's Republic of China, 2015. *Force Development in Critical Security Domains*. Beijing: China Daily. p. 4. Available at http://www.chinadaily.com.cn/china/2015-05/26/content_20820628.htm or http://www.voltairenet.org/article193079.html.

CIA Archives, 2015. *CORONA: Declassified.* Available at https://www.cia.gov/news-information/featured-story-archive/2015-featured-story-archive/corona-declassified.html.

Committee on the Peaceful Uses of Outer Space (COPUOS), 2017. *Guidelines for the Long-Sustainability of Outer Space*, 60th Session, A/AC.105/L.308.

Declaration of Legal Principles Governing the Activities of States in the Exploration and Use of Outer Space, G.A. Res. 1962 (XVIII), 18th Session, Agenda item 28a, A/RES/18/1962 (1963).

Department of State, 1967. *Narrative, Treaty on Principles Governing the Activities of States in the Exploration and Use of Outer Space, Including the Moon and Other Celestial Bodies* (*Outer Space Treaty*). Available at https://www.state.gov/t/isn/5181.htm.

Department of State, 2001. *U.S. Statement, Definition and Delimitation of Outer Space and The Character and Utilization of The Geostationary Orbit, Legal Subcommittee of the United Nations Committee on the Peaceful Uses of Outer Space at its 40th Session in Vienna from April, Statement by the Delegation of the United States of America.* Available at https://www.state.gov/s/l/22718.htm.

Eisel, B., 2005. *The FOBS of War. Air Force Magazine: June 2005.* Available at http://www.airforcemag.com/MagazineArchive/Documents/2005/June%20 2005/0605fobs.pdf.

Gates, R.M., Clapper, J.R., 2011. *National Security Space Strategy Unclassified Summary.* U.S. Government: January 2011, available at: https://www.hsdl.org/?view&did=10828.

Grego, L., 2012. *A History of Anti-satellite Programs.* Union of Concerned Scientists, January 2012. Available at http://www.ucsusa.org/sites/default/files/legacy/assets/documents/nwgs/a-history-of-ASAT-programs_lo-res.pdf.

Harbour, F.V., 1989. The ABM treaty, new technology and the strategic defense initiative. *Journal of Legislation.* 15, 119.

Hickman, J., 2007. Still crazy after four decades: The case for withdrawing from the 1967 Outer Space Treaty. *The Space Review.* September 24, 2007. Available at: http://www.thespacereview.com/article/960/1 (Accessed January 25, 2015).

Kan, S., 2007. *China's Anti-Satellite Weapons Test.* CSR Report for Congress, Order Code RS22652, p. CRS-4, April 23, 2007. Available at https://fas.org/sgp/crs/row/RS22652.pdf.

Kaysen, C., NSAM-192, 1962. *Separate Arms Control Measures for Outer Space*, John F. Kennedy Library and Museum, October 2.

Listner, M.J., 2012. Could commercial space help define and delimitate the boundaries of outer space? *The Space Review.* October 29, 2012. Available at http://www.thespacereview.com/article/2180/1.

Listner, M.J., 2014. Customary international law: A troublesome question for the Code of Conduct? *The Space Review.* April 28, 2014. Available at http://www.thespacereview.com/article/2500/1.

Listner, M.J., 2015. The International Code of Conduct: Comments on changes in the latest draft and post-mortem thoughts. *The Space Review.* October 26, 2015. Available at http://www.thespacereview.com/article/2851/1.

Listner, M.J. Pillai, R.R., 2014. The 2014 PPWT: A new draft but with the same and different problems. *The Space Review*. August 11, 2014. Available at: http://www.thespacereview.com/article/2575/1.

Makarov, A., 2008. Transparency and Confidence-Building Measures: Their Place and Role in Space Security. Security in Space: The Next Generation-Conference Report, March 31–April 1, 2008, United Nations Institute for Disarmament Research (UNIDIR), 2008. See also, UN General Assembly. 43rd Session. 1988. Guidelines for confidence-building measures (A/43/78H).

Mattis, J., 2018. *Summary of the 2018 National Defense Strategy*, January 19, 2018. Available at https://www.defense.gov/Portals/1/Documents/pubs/2018-National-Defense-Strategy-Summary.pdf.

Mineiro, M., 2008. FY-1C and USA-193 ASAT intercepts: An assessment of legal obligations under article IX of the outer space treaty. *Journal of Space Law*. 34.

NASA Policy Instruction, NASA Policy on Planetary Protection Requirements for Human Extraterrestrial Missions, NASA Policy Document 8020.7G. Available from https://www.google.co.uk/url?sa=t&rct=j&q=&esrc=s&source=web&cd=2&ved=0ahUKEwip05TTgIDYAhWlYZoKHUJUBbUQFggvMAE&url=https%3A%2F%2Fnodis3.gsfc.nasa.gov%2FOPD_docs%2FNPI_8020_7_.doc&usg=AOvVaw1Laxcg_3PcvELSWCa89YnE.

National Security Council, NSC-6108, 1961. *Certain Aspects of Missile and Space Programs*. January 18, 1961.

National Science and Technical Council, PDD/NSC 49 (PDD/NSTC 8), 1996. *Fact Sheet, National Space Policy. NASA Historical Reference Collection*. September 19, 1996. Available at http://marshall.wpengine.com/wp-content/uploads/2013/09/PDD-NSC-49-PDD-NSTC-8-National-Space-Policy-19-Sep-1996.pdf.

National Security Planning Board, NSC 5814/1, 1958. *Preliminary U.S. Policy on Outer Space. Dwight D. Eisenhower Presidential Library and Museum*. June 20, 1958, p. 7. Available at http://marshall.wpengine.com/wp-content/uploads/2013/09/NSC-5814-Preliminary-U.S.-Policy-on-Outer-Space-18-Aug-1958.pdf.

Obama, B., 2010. *National Space Policy*. June 28, 2010. Available at http://marshall.wpengine.com/wp-content/uploads/2013/09/Fact-Sheet-The-National-Space-Policy-June-28-2010.pdf.

Office of the Historian, Department of State, 1977. *Memorandum from the President's Assistant for National Security Affairs (Brzezinski) to Secretary of State Vance, Secretary of Defense Brown, the Director of the Office of Management and Budget (McIntyre), the Director of the Arms Control and Disarmament Agency (Warnke), the Chairman of the Joint Chiefs of Staff (Brown), the Director of Central Intelligence (Turner), the Administrator of the National Aeronautics and Space Administration (Frosch), and the Special Advisor to the President for Science and Technology (Press)*. September 23, 1977. Available at https://history.state.gov/historicaldocuments/frus1977-80v26/d11.

Office of the Press Secretary, NSD-30 (NSPD-1), 1989. *National Space Policy*. November 16, 1989. Available at http://marshall.wpengine.com/wp-content/uploads/2013/09/NSD-30-NSPD-1-National-Space-Policy-November-2-1989.pdf.

Paine, T., 2015. Bombs in Orbit? Verification and violation under the Outer Space Treaty. *The Space Review*, March 18, 2015. Available at http://thespacereview.com/article/3454/1.

Parkenson, J.E., 1987. International legal implications of the strategic defense initiative. *116 Military Law Review*. 67.

Preston, R., Johnson, D.J., Edwards, S.J.A., Miller, M., Shipbaugh, C., 2002. *Space Weapons Earth Wars*. RAND. pp. 36–49. Available at https://www.rand.org/content/dam/rand/pubs/monograph_reports/2011/RAND_MR1209.pdf.

Reagan, R., NSDD 42, 1982. *National Space Policy*. July 4, 1982, para. IV(B). Available at https://www.reaganlibrary.archives.gov/archives/reference/Scanned%20NSDDS/NSDD42.pdf.

Reagan, R., NSDD-172, 1985. *Presenting the Strategic Defense Initiative*. May 30, 1985. Available at https://www.reaganlibrary.archives.gov/archives/reference/Scanned%20NSDDS/NSDD172.pdf.

Reagan, R., NSSD 4-86, 1986. *Anti-Satellite (ASAT) Options*. October 20, 1986. Available at https://reaganlibrary.archives.gov/archives/reference/Scanned%20NSSDs/NSSD4-86.pdf.

Reisman, M.W., Armstrong, A., 2006. The past and future claim of preemptive self-defense. *The American Journal of International Law* 100.3 (2006), 525–550.

Scowcroft, B., 1977. NSDM 345. U.S. Anti-Satellite Capabilities. Gerald. R. Ford Library, January 18, 1977.

Sellers, J.J., 2004. *Understanding Space: An Introduction to Astronautics*, 2nd ed. Boston: McGraw Hill.

Shixiu, B., 2007. Deterrence revisited: Outer space. China Security World Security Institute. Winter 2007 (for a perspective on the United States 2006 National Space Policy and its effect the PRC's national security). 2–12. Available at https://www.yumpu.com/en/document/view/26458606/china-security-space-library/6.

Stone, C., 2015. Security through vulnerability? The false deterrence of the National Security Space Strategy. *The Space Review*. April 15, 2015. Available at http://www.thespacereview.com/article/2731/1.

Terrill, Jr. D.R., 1999. *The Air Force Role in Developing International Space Law*. Air University Press, Maxwell Air Force Base, Alabama.

Trump, D.J., 2017. *National Security Strategy of the United States of America*, December 2017. Available at https://www.whitehouse.gov/wpcontent/uploads/2017/12/NSS-Final-12-18-2017-0905.pdf.

Tzu, S., 2017. *The Art of War (forward by James Clavell)*. New York: Hodder.

Union of Soviet Socialist Republics: Draft Declaration of the Basic Principles Governing the Activities of States in Pertaining to the Exploration and Use of Outer Space, A/AC. 105/L.2 (Distr. LIMITED) September 10, 1962.

United States Army, 1959. Project Horizon Report, A U.S. Army Study for the Establishment of a Lunar Outpost, Volume I, Summary and Supporting Considerations, June 9, 1959 [Regraded Unclassified, September 21, 1961], p. 1, para. 3(a)(1) and 3(b)(1).

von der Dunk, F., 1998. The Delimitation of Outer Space Revisited: The Role of National Space Laws in the Delimitation Issue. *Space and Telecommunications Law Program*, Faculty Publications, January 1, 1998. Available at http://digitalcommons.unl.edu/cgi/viewcontent.cgi?article=1050&context=spacelaw.

Weeden, B., 2014. *Through a glass, darkly: Chinese, American, and Russian anti-satellite testing in space*. Secure World Foundation: March 17, 2014. Available at: https://swfound.org/media/167224/through_a_glass_darkly_march2014.pdf.

Williamson, M., 2010. *A Pragmatic Approach to the "Harmful Contamination" Concept in Article IX of the Outer Space Treaty*. 5th Eileen M Galloway Symposium on Critical Issues in Space Law. December 2010. Available at: http://www.spacelaw.olemiss.edu/events/pdfs/2010/galloway-williamson-paper-2010.pdf.

Wilson, H., Goldfein, D., Raymond, J., Greves, S., 2017. Department of the Air Force, *Presentation to the Subcommittee on Strategic Forces, United States Senate, Military Space Policy*, May 17, 2017, p. 2. Available at https://www.armed-services.senate.gov/imo/media/doc/Wilson-Goldfein-Raymond-Greaves_05-17-17.pdf.

CHAPTER 10

Managing the Resource Revolution

Space Law in the New Space Age

Thomas Cheney and Christopher J. Newman

CONTENTS

10.1 INTRODUCTION

The rise of human space activity has seen a number of risks and challenges develop from current technology and engineering capabilities. The emergence of human-made orbital debris and the risks posed by state and extremist group violence are all indicative of the new frontiers of risk, which are opening up due to a significant expansion in private and state-based space capabilities. One area of risk yet to manifest itself, but one which is clearly in the contemplation of both policy makers and strategists comes from an area yet to see any space activity; that of the mining and exploitation of celestial bodies, such as the Moon and Near-Earth Objects (NEOs). Unlike other areas of human space activity, the risks at this stage are purely speculative, given that the necessary technological infrastructure does not yet exist to even begin exploratory mining ventures.

The chapter will investigate this new frontier of risk and explore how the mere prospect of asteroid mining threatens to unbalance the current legal and political cooperation which regulates space activity. In many respects, this chapter will be diametrically opposed to other inquiries into the viability of space mining. Traditional treatises on space mining tend to brim with optimism on the untold riches of space, which are just waiting to be claimed once the technology is ready. This discussion, however, will seek to emphasize the prosaic, terrestrial difficulties that any new space mining enterprise faces. The lack of legal certainty over property mined in outer space means that investors are hesitant to invest large sums of capital where there is no guarantee that minerals would be freely tradable. This lack of investment inhibits research and development activity and makes the prospect of space mining ever more distant. Without a certain and internationally legitimate legal framework, investment in space mining is likely to stagnate and attempts to unilaterally legislate have already caused international tension and pose a serious risk to the stability of space governance.

The discussion will start by exploring why space mining poses such a challenge to current space law and policy. It will identify the potential riches and outline the numerous commercial actors actively engaged in exploring this area. There will then be an assessment of the technical issues that have emerged and need to be addressed before any significant commercially viable enterprises can be established. Following on from this, the aforementioned legal and policy regime will be critically examined. At present, it is not technical difficulties but legal and regulatory issues regarding the distribution and ownership of the mined resources that present the clearest danger to stable and sustainable space activity. Any discussion of future space mining ventures overlooks these issues at its peril.

By focusing on the legal and policy debates, the chapter will forsake speculation as to the infrastructure and engineering solutions, which have yet to be truly conceptualized, let alone developed into working systems. Indeed, it is something of a paradox that the biggest risk posed by this future space activity is to cooperation and governance arrangements in the present day. The fracturing of this consensus, over disputes as to how to allocate and manage resources, could place at risk the creation of the very infrastructure needed to facilitate asteroid mining. This chapter will ultimately evaluate both the causes for pessimism and optimism in this area and attempt to outline the way in which the legal risks can be mitigated, while projecting policy solutions to foster a resource utilization regime accepted by all stakeholders.

10.2 SPACE MINING IN CONTEXT

At first sight, the economic case for investing in asteroid mining appears to be a compelling one. It would appear that there are substantial amounts of precious and valuable metals in asteroids scattered throughout the Solar System. It has been suggested that Amun, a small NEO with a mass of approximately 30 billion tons, contains approximately $8000 billion in iron and nickel, $6000 billion in cobalt, and $8000 billion in platinum group metals. Similar estimates have projected that the asteroid belt also contains about 4 billion tons of uranium (Lewis 1997). While the Moon and other planets may have even more lucrative resources, asteroids, and in particular NEOs have the added lure of being "the most easily reachable bodies within the entire Solar System." There are estimated to be 20,000 NEOs larger than 100 m diameter, all capable of being mined in the near future given sufficient investment (Di Martino et al. 2009 at p. 195).

As well as their relative convenience and abundance of minerals, another aspect of asteroids and NEOs, which makes them attractive propositions for mining ventures, is the potential to utilize the water present on such bodies (Lewis 2015). Water is a valuable commodity in space; it can be used for drinking, bathing, and cleaning but it can also be used to make air and rocket fuel. As it costs around $20,000 to put a typical 500 ml bottle of water into orbit it would be vastly more efficient and cost-effective to use a space-based source of water rather than rely on a supply from Earth (Shepard 2015). Asteroid mining for water ice is technologically feasible and would be achievable using established technology (Lewis 1999).

The production of fuel in space would be of interest to a number of stakeholders in the space sector. First, the mining industry itself would more quickly be able to become self-sustaining by employing such in situ resources. Additionally, on-orbit servicing is, much like the space mining sector, a developing and embryonic industry, which would greatly benefit from a comparatively cheap source of fuel (Henry 2017). Finally, established space companies such as the United Launch Alliance (ULA) have indicated that they would be willing to pay $3000 for a kilogram of propellant delivered to low Earth orbit (LEO) (David 2016). This projection fits well with the assessment made by Lewis, that delivery to Earth orbit for less than $10,000 per kilogram would be competitive with Earth-launched material (Lewis 1999 at p. 113). In the future, it is not difficult to envisage the creation of a series of space-based "filling stations" processing locally sourced water and facilitating travel into the Solar System.

10.2.1 Nascent Space Mining Industry

Even discounting the lucrative mineral deposits, the exploitation of asteroids and NEOs as part of a wider infrastructure to support deep space exploration becomes a very attractive proposition. In respect of convenience and accessibility, the Moon is also attracting the attention of commercial enterprises. Moon Express and iSpace are both companies having invested in research and development of technology capable of exploiting lunar resources. Those advocating for the exploitation of the Moon have tended to focus on the presence of helium-3, an extremely rare isotope on Earth, which has potential to be a fuel source for fusion reactors (Bilder 2009–2010). The utility of helium-3 mining is, however, contingent upon technology becoming viable for widespread usage and thus at present this prospect is somewhat remote (Newman 2015 at p. 31). There are, however, other materials, such as rare earth metals and water ice, which make the Moon of interest to developers. This is especially the case if the resources were in support of a manufacturing or servicing industry in LEO, supporting lunar bases and/or a developing lunar economy (David 2015). At present, such discussions may seem somewhat far-fetched, yet the proposals for a Moon village from ESA and commercial "space hotels" from Bigelow Aerospace illustrate that such ideas could soon emerge as serious propositions (Woerner 2016).

Given the apparent abundance of minerals and the ability to use in situ materials to obtain these resources, it was inevitable that companies would emerge and try to exploit these opportunities. Two companies, Planetary Resources and Deep Space Industries, announced their intentions to commence commercial asteroid mining within the near future in April 2012 and January 2013, respectively. Their proposals utilize current technology and their business plans, while ambitious, do lay out a convincing method of creating a profitable enterprise (Foust 2012). So convincing were their proposals that they have successfully garnered support from the government of Luxembourg. In both the United States and Luxembourg, this support has also been backed by legislation. The United States has Title IV of the U.S. Commercial Space Launch Competitiveness Act* and Luxembourg, the Law on the Exploration and Use of Space Resources.† Both the American and Luxembourg laws are drafted so as to be able to cover space mining of any and all celestial bodies, the focus is on resources not the "territory" they are found on or in and as such could apply to the Moon or asteroids alike.

10.2.2 Introducing the Legal Dimension of Space Mining

There are significant legal and policy issues, which need to be addressed if (to paraphrase one headline) space mining is to take a giant leap from science fiction to science fact (Cookson 2017). The promulgation of national legislation creating a process for the authorization of space mining has prompted discussion at the international level, particularly the Legal Subcommittee of the United Nations Committee on the Peaceful Uses of Outer Space (COPUOS), where it has been discussed at both the 2016 and 2017 sessions and is on the agenda for the 2018 session as well. It has also been the focus of The Hague Space Resources Governance Working Group, a multinational interdisciplinary group comprised of experts and other stakeholders. The Hague Working Group has at the conclusion of Phase One of their work released a set of "Draft Building Blocks for the Development of an International Framework on Space Resource Activities." This seeks to initiate discussion about creating an international consensus for the governance of space resource activities.

---

* Available at https://www.congress.gov/114/plaws/publ90/PLAW-114publ90.pdf.

† Available at http://data.legilux.public.lu/file/eli-etat-leg-loi-2017-07-20-a674-jo-fr-pdf.pdf.

Despite the promulgation of these national laws and discussions on an international level, there are significant legal obstacles preventing the establishment of a broad consensus at present. These will be considered in some detail later. The discussion will at first turn to consider the technical and engineering aspects of space mining. The technology is developing as the industry is developing, with investment and research and development looking to overcome the formidable challenges facing those who seek to create the necessary infrastructure. Consideration of the technical difficulties will help place the legal and policy solutions in context, and assist in the broader aim of creating a robust and enduring governance framework.

10.3 THE TECHNICAL ASPECT OF SPACE MINING

It is axiomatic to say that the space mining industry is still in the early phases of its development. It has yet to engage in rudimentary operational testing, much less conduct mining activity. The developmental nature of the hardware means that the final picture of how space mining will be technically accomplished is still unclear. Additionally, understanding of the actual makeup of celestial bodies is limited. Sample return missions have been conducted, with the United States, the Soviet Union, and Japan having all returned samples from celestial bodies to Earth, primarily the moon but also asteroids and comets (Willis 2016). This means that there exists some experience with extracting material from celestial bodies and returning them to Earth; however, it is a weak foundation upon which to establish a regulatory regime.

10.3.1 The Composition of Celestial Bodies

One of the major issues facing the space mining industry is the lack of scientific knowledge of the makeup of asteroids and comets. This was partially demonstrated by the *Rosetta* mission, which required multiple forms of landing methods because the mission planners only had a limited idea of what they would find in respect of the comet surface. Furthermore, the actual make up of asteroids and comets could pose an obstacle to commercial mining operations. The current knowledge base is limited for a foundation upon which to base commercial operational planning. The accuracy of astronomical observations needs to improve in order to be useful for commercial operations; the current standard of accuracy is limited—which is less important for scientific endeavors—yet remains a significant obstacle for companies who will not want to waste time or money on exploration of noncommercially viable objects (Elvis 2013). Current optical-near-infrared spectra can only tell us the mineral content of the asteroid's surface though for most, if not all asteroids, surface content should be representative of the whole object (Elvis 2013 at p. 91). Spectrometric observations are not reliable to a commercial standard, for example, M-class asteroids were thought to be primarily Iron (Fe) and Nickel (Ni) but it turns out that they have much more silicate content than was thought (Granvick et al. 2013 at p. 151).

This shortfall in compositional knowledge has consequences for any potential legal categorization of celestial bodies too. Asteroids are defined by their composition—not their size—and the current state of understating of their composition makes it unwise to use the existing knowledge bases for the establishment of a legal categorization system. Additionally, astronomical terms themselves are vague and "any small sized body orbiting the Sun could be defined an asteroid" (Di Martino et al. 2009 at p. 72). The difference between asteroids and comets is not only somewhat vague but also transitory, as over time comets have their volatiles "baked off" and come to resemble asteroid-like objects (Di Martino et al. 2009 at p. 73). While it is true that there are thousands of NEOs larger than 100 m diameter and vast numbers of even smaller NEOs, it may prove to be a challenge to find commercially viable asteroids as it is necessary to find materials commercially viable to mine. Indeed, it could be that "finding suitable asteroids to mine could well be the bottleneck to developing asteroid resources" (Elvis 2013).

In summary, and as previously identified, there is still a substantial amount of research and development to undertake on the hardware and yet further scientific discovery as to the nature of celestial bodies before space mining can move beyond this nascent phase of planning and into actual space activity. These technical difficulties may well be overcome given sufficient investment and attention. Having considered these areas, the discussion will move on to consider in more depth the legal and policy problems regarding ownership and regulation of mining. In order to fully appreciate the depth of these difficulties and the risk posed to wider cooperation in space, it is necessary to understand the origins of the laws governing space activity. Consideration of the foundations of the existing legal framework, with special regard to property ownership in space, will illustrate the conceptual and attitudinal shift required in order to create a meaningful set of property rights for the fruits of space mining.

10.4 THE EXISTING LEGAL FRAMEWORK FOR OUTER SPACE ACTIVITY

At first sight, it may seem somewhat hyperbolic to talk about the legality of property ownership as being a significant barrier to the establishment of an economically viable space mining industry. Yet, one of the underpinning precepts of international space law is that outer space and celestial bodies should not be subject to claims of ownership or sovereignty. This provision arose due to the conflict of the cold war, the backdrop against which space law was formed (Blount 2012). Yet despite the changing geopolitical situation, the resulting treaty, known colloquially as the Outer Space Treaty (OST) of 1967 remains the cornerstone of the international space law regime. There are additional treaties, such as the Moon Agreement,* that also have relevance for the topic of space mining, as well as the UN General Assembly resolutions, customary international law, and so-called "soft law" instruments, which expand upon and develop the principles outlined in the OST. Yet it is the 1967 treaty which remains the focal point and has been described as the "constitution" of outer space (Gangale 2009 at p. 52). It enjoys virtually universal recognition having been ratified by 107 states and signed by a further 23 including all the space capable states. Many of the provisions of the OST may well be considered to be customary international law, most specifically Articles I-IV and VI. An argument can be made for the rest of the provisions of the treaty having also achieved that status, given its broad acceptance (Larsen 2014 at p. 289).

10.4.1 The OST and Space Mining

Regarding space mining, the key provisions of the OST are Articles I-III and VI. Articles I and II are often read together and are certainly complimentary (Lyall and Larsen 2009 at p. 180). Article I is a statement of the basic principle of the freedom of exploration, access, and use of outer space for all countries. Article II of the OST codifies the nonappropriation principle, which was established in UN General Assembly resolution 1721.† It establishes that "outer space, including the Moon and other celestial bodies, is not subject to national appropriation by claim of sovereignty, by means of use or occupation, or by any other means." This was one of the earliest and most widely agreed principles of space law (Freeland and Jakhu 2009).

* Agreement Governing the Activities of States on the Moon and Other Celestial Bodies (adopted December 18, 1979, entered into force July 11, 1984) 1363 UNTS 3 (Moon Agreement/MA).

† UNGA Res 1721 (December 20, 1961) UN Doc A/RES/1721 (XVI), A1(b).

Articles I and II work in conjunction and are what make space *res communis* (a thing belonging to the entire community) rather than *res nullius* (belonging to no one). Article III of the OST establishes that there is the rule of law in outer space and declares that space activities shall be carried out in accordance with international law "including the Charter of the United Nations, in the interest of maintaining international peace and security and promoting international cooperation and understanding."

If the first five articles of the 1967 treaty can be said to establish broad behavioral norms, Articles VI, VII, and VIII of the OST lays down the contours of state responsibility. Article VII establishes state liability for damage caused by space objects and Article VIII provides that ongoing jurisdiction and control of space vehicles and personnel remains with the state on whose registry the craft is carried. Most crucially for space mining, Article VI articulates the key principle that states are responsible for the actions of their nationals in outer space.

Under Article VI, governments are required to authorize and continually supervise their own national space activities. This is usually achieved by the enactment of primary national space legislation such as the United Kingdom's Outer Space Act of 1986 or Australia's Space Activities Act of 1998. It is through Article VI of the OST that the nonappropriation principle found in Article II of the OST can be said to apply to corporations and natural persons (referred to in the treaty as nongovernmental entities) as well as states themselves (Tronchetti 2015a).

10.4.2 The Moon Treaty in Context

Beyond the OST, there are several other international space treaties, among which the most relevant for this discussion is the Moon Agreement. The Moon Agreement is generally regarded as being a "failed treaty" (von der Dunk 2012) due to its low uptake of 17 ratifications and 4 signatures.* It has, nonetheless, achieved the requisite number of ratifications under the UN rules and is a valid active treaty. Indeed, it is still gaining new parties; Venezuela became a party to the treaty as recently as November 2016.† For the most part, the Moon Agreement replicates the OST, however regarding the topic of space mining Article 11 is the most relevant. Article 11 largely expands upon the provisions of Article II of the OST.

* http://www.unoosa.org/documents/pdf/spacelaw/treatystatus/AC105_C2_2017_CRP07E.pdf.

† UNOOSA, "Accession by Venezuela (Bolivarian Republic of) to the Agreement Governing the Activities of States on the Moon and Other Celestial Bodies" (November 3, 2016) UN Doc C.N.829.2016.TREATIES-XXIV.2 (Depositary Notification).

Yet Article 11(1) provides that "the Moon and its natural resources are the common heritage of mankind;" this is the infamous "common heritage of mankind" principle, which can also be found as a distinct concept in the Law of the Sea Convention (UNCLOS).* Accompanying this, is the provision in Article 11(5) of the Moon Agreement for the establishment of an international regime to "govern the exploitation of the natural resources of the Moon..." and the provision in Article 11(7)(d) for "an equitable sharing by all States Parties in the benefits derived from those resources..." This common heritage principle has attracted significant criticism in respect to both uncertainty of its scope (Jakhu 2005) and the different interpretations which could be placed on the management of resources by individual states (Tronchetti 2015a).

The basic legal framework governing space activities has remained largely unaltered from the cold war era. The OST sought to preserve peace in space, as this was its primary purpose. Issues such as space mining were, however, not within the sensible contemplation of the drafters of the treaty, preoccupied as they were with the race to the Moon. Nonetheless, the principles that found their expression in the 1967 treaty have become normative values, accepted by all members of the spacefaring community. This, however, causes substantial problems for space mining in three key respects. First, Articles I and II of the OST have placed a significant question mark over whether space resources can ever be "owned." Second, Article VI means that states are responsible for ensuring all of their national space activities are compliant with the treaty, but authorizing space mining in a manner compliant with Article II is extremely problematic. Finally, the inclusion of Article 11 is undoubtedly the reason why the Moon Treaty has not received broader acceptance and remains largely ignored by spacefaring nations. Yet Article 11 of the 1979 treaty was, perhaps, the best chance of establishing an international consensus on the way space resources could be administered and managed. Having outlined these basic principles of space law, the discussion will now directly address the legality of space mining.

10.5 THE LEGALITY OF SPACE MINING

This question of whether space mining is permitted under the existing treaty regime has gained significant traction as more companies, looking for investment, seek a more certain legal framework upon which to base

* http://www.un.org/depts/los/convention_agreements/texts/unclos/unclos_e.pdf, Article 136.

their plans. As stated, Article II prohibits states from owning territory on the Moon or any of the planets or celestial bodies; this is a broad prohibition on appropriation, which applies to their nationals and bodies corporate via Article VI of the treaty. It is not clear whether this prohibition extends to natural resources contained on or in these celestial bodies and the Treaty is silent on this aspect. On a strict reading of the terms of the treaty, it has been argued that Article II does indeed prohibit such exploitation and that natural resources cannot, legally speaking, be separated from the celestial body in which they are found (Tronchetti 2015a at p. 790). There is now, however, an emerging body of opinion taking the view that resources and celestial bodies can be legally distinct, specifically after the resource has been extracted. An analogy would be fish in the high seas: no one can appropriate the high seas but fishermen who extract the fish from the seas can appropriate those fish and lawfully trade them as commodities (Dula and Zheniun 2015).

10.5.1 Domestic Space Resource Legislation

The above view taken by the United States since the enactment of the U.S. Space Resource Exploration and Utilization Act of 2015,* and Luxembourg in their respective space mining law. Indeed, the Luxembourg legislation refers specifically to this in the explanatory document published alongside their draft space mining law.† Furthermore, there is state practice supporting this argument in that several governments (the United States, the USSR, and Japan) have conducted scientific sample return missions to various bodies in the Solar System and all claim ownership of their respective samples. Indeed, the Russian Federation has even sold portions of the Soviet lunar samples at an auction in 1993 without objection from the international community (Pop 2013). Whether this constitutes sufficient state practice and *opinio juris* to create a customary rule is certainly a question worthy of consideration (Listner 2014).

Somewhat unhelpfully, the term "celestial bodies" is used frequently throughout space law instruments, yet there is no definition of the term provided by the OST or any of the other space treaties. This is odd for a UN treaty of a general nature like the OST (Hobe 2009 at p. 29). Some have

* U.S. Commercial Space Launch Competitiveness Act, Public Law 114-90, 114th Congress, November 25, 2015, 51 U.S.C., Title IV.

† Luxembourg Ministry of the Economy (2016) "Draft Law on the Exploration and Use of Space Resources." Available at: http://www.gouvernement.lu/6481974/Draft-law-space_press.pdf (Accessed March 12, 2017).

argued that this could potentially affect which naturally occurring "space objects" are subject to the terms of the OST, and specifically which fall under the prohibition on national appropriation laid out in Article II of the OST (Pop 2009). If a naturally occurring space object is not a celestial body then it may not fall under that prohibition. If asteroids, or even certain asteroids are, for example, not celestial bodies—at least in the legal sense as meant by the treaties—then they would be free for appropriation. This would of course depend on what the definition of a celestial body is.

10.5.2 Defining "Celestial Body"

In his book, *Who Owns the Moon,* Virgiliu Pop has argued that there are four approaches to defining celestial bodies: (1) the "spatialist approach," categorizing naturally occurring objects based on their size; (2) the "control approach," categorizing an object based on the ability of humans to move it; (3) the "functionalist approach," differentiating between objects treated as celestial bodies and those simply being used as moveable ore bodies; and (4) the "space object" approach, arising out of the discussion of the possibility of converting asteroids into spaceships, allowing for converted asteroids to be registered as "space objects" (Pop 2009 at pp. 51–57). Pop's four approaches can essentially be reduced to two approaches, which provide three options for legal classification; either asteroids and comets are categorized by their size or by their ability to be moved by artificial means. A third option is to state that all are celestial bodies and neither their size nor our ability to move them makes any difference to their legal status. This approach is the one that most closely accords with the interpretive rules laid out in Article 31(1) of the Vienna Convention on the Law of Treaties and has been endorsed by a number of academics and international bodies.*

There is a further complication with regard to those few states that are parties to the Moon Agreement. There is a strong argument to be made that the contentious Article 11 of the Moon Agreement places a moratorium on mining operations until the "international regime" is established to govern those activities. This argument is strengthened by the fact that states are not meant to undermine the object and purpose of a treaty they are a party or signatory to, which the unilateral authorization of space mining prior

* The Space Generation Advisory Council's (SGAC) Space Law and Policy Project Group took this view in their space mining position paper, and is a view which was endorsed by Mark Sundahl in Sundahl, M. "Don't muddy the message to space mining companies" Space News available at: http://spacenews.com/op-ed-dont-muddy-the-message-to-space-mining-companies/ (accessed November 2017).

to the establishment of the required international regime would do (Jakhu et al. 2013). Obviously, this has little impact upon those states that are not parties to the Moon Agreement but it is a further complication of the legality of space mining, which is worth considering, and is potentially a source of diverging space law regimes between states based on whether they are party to the Moon Agreement or not.

International space law is inherently a permissive and voluntary system, meaning states retain freedom of action unless specifically prohibited. Furthermore, Article I of the OST provides a very broad freedom of use of space. There is no definition of use given, therefore, in accordance with the rules of treaty interpretation laid out in the aforementioned Vienna Convention on the Law of Treaties it is the "plain ordinary meaning" of the term "use" that should be used in interpreting Article I of the OST. The *Oxford English Dictionary* defines use as to "take, hold or deploy as a means of achieving something or take or consume (an amount) from a limited supply," which would therefore mean that space mining could be included within the freedom of use laid out in Article I of the OST. This is of course tempered by the prohibition on national appropriation in Article II of the OST—and indeed the rest of the provisions of the treaty including other provisions of relevant international law—but it does not prohibit space mining, though it will have implications for how it is regulated, authorized, and conducted.

10.6 A RISK TO THE INTERNATIONAL ORDER: 21ST CENTURY FRACTURING OF THE CONSENSUS?

As previously discussed, states have an obligation under Article VI of the OST to "authorize and supervise" the activities of their nationals in outer space. To this end most states with active private space sectors have national legislation providing for a process of "authorization and supervision." The United States and Luxembourg both felt that their existing legislation was inadequate for the task of "authorizing and supervising" space mining activities and therefore chose to introduce national space mining laws in response to the development of a space mining industry in their jurisdictions.

In November of 2015, the United States passed the Commercial Space Launch Competitiveness Act,* Title IV of which dealt with Space Resource

* U.S. Commercial Space Launch Competitiveness Act, Public Law 114-90, 114th Congress, November 25, 2015, 51 U.S.C.

Exploration and Utilization. It defined an "asteroid resource" as "a space resource found on or within a single asteroid"* and a space resource as "an abiotic resource in situ in outer space,"† which "includes water and minerals."‡ It also stipulates that a U.S. citizen is entitled to "any asteroid resource or space resource obtained, including to possess, own, transport, use, and sell the asteroid resource or space resource obtained in accordance with the applicable law, including the international obligations of the United States."§ Luxembourg followed suit in 2017 with a law of its own, which—while more detailed than that of the United States—largely mirrors it with a few notable exceptions. Luxembourg will only grant authorization to legal persons incorporated in Luxembourg (and only to the two specific types of corporation stipulated in the law, roughly analogous to the limited liability corporations found in the United States and United Kingdom), and there are criminal penalties stipulated for noncompliance.¶ There are other states expected to follow with space mining laws of their own in the next few years.

The OST is now 50 years old; the governance regime it established has served space well. Space has become a vital part of Earth's infrastructure and economy, which has been made possible by the order and stability provided by the space law regime that rests upon the foundation provided by the OST (see Johnson-Freese 2017). However, space mining has the potential to undermine the stability of the space law regime. There are three main potential friction points, which will be discussed below. The first potential conflict is over the legality of space mining itself, which has been already touched upon, with a second potential area of conflict being over the actual resources being mined, while the third area of conflict is over the distribution of the profits from space mining. These three issues have the potential to destabilize or delegitimize the space law regime without which the economic value of space would considerably diminish.

10.6.1 Risks from an Uncertain Legal Framework

The first potential area for conflict or crisis in space law is over the legality of space mining itself. The 2016 Legal Subcommittee of COPUOS saw several delegations, most notably that of the Russian Federation, strongly

* §51301(1).
† §51301(2)(A).
‡ §51301(2)(B).
§ §51303.
¶ http://data.legilux.public.lu/file/eli-etat-leg-loi-2017-07-20-a674-jo-fr-pdf.pdf.

object to and criticize the U.S. Space Resource Exploration and Utilization Act of 2015, focusing on the perceived unilateral nature of the U.S. space law (UNCOPUOS 2016). This has not deterred the United States or Luxembourg from enacting the space mining laws.

There are essentially three arguments against the legality of space mining and the U.S. and Luxembourg position in particular. The first is, as has already been mentioned, that Article II of the OST creates a total prohibition on property rights in space and this includes commercial space mining operations. This involves a strict interpretation of the definition of appropriation. Those who articulate this viewpoint also generally see resources as being inseparable from the celestial body they are found in; as appropriation of celestial bodies is prohibited in whole or in part then so is the appropriation of their resources (Larsen 2014). The second is that these national laws are an act of sovereignty and are therefore incompatible with the space law regime. The third is that space mining can only be legal under an international regime.

The view that was articulated by several states at the 2015 session of the Legal Subcommittee of the UN Committee on the Peaceful Uses of Outer Space is that the U.S. space mining law constitutes an act of national appropriation, which is incompatible with the OST—Article II in particular. Authorizing mining operations and/or granting title over extracted resources is an act of national appropriation, which is in violation of Article II. In order for a government to have the authority to regulate an activity, they need to have jurisdiction over the area where the activity is being conducted. By claiming jurisdiction over the celestial body the state violates Article II of the 1967 treaty. The government in this case is granting title to the extracted resource and is thus claiming ownership by virtue of authorizing the transfer of ownership (Tronchetti 2015b).

This approach suggests that space mining can only be compliant with international space law if it is sanctioned by an international regime. If space belongs to the international community, it is part of the global commons and therefore no individual state has the right to authorize its nationals to conduct mining within it. Only the international community working together, preferably through the United Nations, can offer legal consensus on space mining. This holds even truer for those states that are party to the Moon Agreement, who look toward Article 11 of that treaty (see UNCOPUOS 2016).

The counterargument to these viewpoints is that the U.S. and Luxembourg positions are a legitimate interpretation of the OST and it is within the rights of a "State Party" to unilaterally interpret their obligations under a treaty. It is within the purview of states to authorize and supervise space mining through the means of national legislation. Both the United States and Luxembourg make no claim to territorial sovereignty or control over any celestial body either in whole or in part, or indeed to the resources in situ. They are regulating the activities of their nationals, as they are required to do by Article VI of the OST. Furthermore, they reject the requirement for an international regime as neither are party to the Moon Agreement and therefore are not bound by it (UNCOPUOS 2016). This approach was endorsed by the International Institute of Space Law—as being a valid interpretation of international law—not long after the promulgation of the U.S. space mining law.

10.6.2 The Risk of a Resource Stampede

The risk of these diverging opinions is of the emergence of two distinct power blocs. The first bloc would regard space mining as legal and legitimate and would feel it needs to circumvent the UN system in order to effectively manage the resources emerging from it. The other power bloc would regard either space mining or the national legal regimes underpinning it as incompatible with international law and as such, would consider the fruits of mining as being illegitimate. The development of this schism has the potential to undermine the generally collaborative space law framework. Given the nature of space, this collaboration undoubtedly needs an internationally recognized and respected framework in order to continue; otherwise the ripples of discord could be felt in all areas of space activity (see, e.g., governance of the radio spectrum). A breakdown in the established space law regime could prove seriously detrimental to the value of the space environment for all actors.

A further risk to peace, and a potential source of conflict, is over the resources themselves. It is becoming clear, even from limited scientific data, that there is an abundance of resources in the Solar System. Where are those resources located and how easy are they to access is not as clear. The initial target for space mining operations will be the Near-Earth Objects (NEOs), as their location makes them relatively easy to access. As stated above, much is still unknown about the distribution of resources among NEOs. If easy to access resource-rich asteroids turn out to be a rarity, then that could cause problems (Shepard 2015). The limitations of asteroid mining in this respect are particularly important when considering the concerns of developing nations, particularly regarding equitable access.

If it does turn out that the number of commercial viable asteroids and other NEOs are limited, then their concerns could further lead to fractures in the space governance framework and have wider diplomatic ramifications.

Articles I and II of the OST make space a *res communis*, part of the global commons. Furthermore, the preamble of the OST declares that activities should be carried out for the "benefit of all peoples," which is reaffirmed in Article I. Traditionally, large-scale space activity has excluded the "global south" and it is argued there needs to be some kind of benefit sharing arrangement in the event of space mining. The fear is of a repeat of colonialism in space where Western or developed states reap the rewards and afterward there is either little left (by the time the lesser developed states get into space) or what is left is harder to access (Oduntan 2005).

10.6.3 Potential Disputes over Profit Sharing

It was these fears, regarding the potential of both deep-sea mining and space mining, which drove the development of the "common heritage of mankind" principle found in the UNCLOS and the Moon Agreement (Christol 1981). There are still calls for equitable sharing of space benefits although they do not have the same intensity as they did during the era of the "New International Economic Order." Glimpses of this conflict are already being seen, however, if substantial and lucrative mining operations begin they could further fuel resentment and tension. There have already been suggestions of this in other areas of space governance such as the Bogotá Declaration,* in which equatorial states declared ownership over the geostationary orbit located above the equator—and thus above their countries. They felt that the developed world was unfairly reaping the benefits of the geostationary orbit and wanted "a slice of the pie." These are concerns that will continue to manifest themselves if access to space is seen as the preserve of the established space powers (Zullo 2001–2002).

Given the cost and complexity of accessing space it seems unlikely that widespread armed conflict will happen in space, although this may change soon, at least beyond LEO. It is not unimaginable, given the vast potential wealth available, that this might occur in the future should perceived inequality of access continue. There is precedent from history—with organizations like the East India Companies having considerable naval capacity to protect their wealth—and even today with the terrestrial

* http://www.spacelaw.olemiss.edu/library/space/International_Agreements/declarations/1976_bogota_declaration.pdf.

mining industry employing private military contractors to protect their investments in the more dangerous areas of the developing world. However, conflict over resources is much more likely to take the form of legal and diplomatic conflict than the armed variety. This will all have a cost and has the potential to undermine the legitimacy and effectiveness of the overall space law regime, especially if the existing system is unable to satisfactorily resolve disputes.

The U.S. and Luxembourg space mining laws only apply to their respective nationals. This is not necessarily a problem if the space mining industry is restricted to a handful of actors (and at the moment the embryonic industry is dominated almost to the point of exclusivity by American-based companies), but combined with the lack of a dispute resolution system in space law this could be a potential source of conflict were two companies from different states try to mine the same asteroid. There is the provision in Article IX of the OST against "harmful interference" with another state's space activities; however, neither what constitutes "harmful interference" nor who is responsible for preventing it have ever been clarified.

Additionally, there are also concerns about the effect that space mining of metals would have on the terrestrial markets as many developing states rely on mining as an important source of income. Neither of the two main asteroid mining contenders are discussing either mining metals or bringing them to Earth in the near future (Cookson 2017). It also does not at present (or any time in the near future) make economic sense to return material to Earth (Ryan and Kutschera 2013). That does not stop it from being a source of concern or opposition, and is thus something which needs to be discussed (UNCOPUOS 2016).

Clearly, however, as identified above the main risk to the Earth from space mining comes from an inherently damaging fracture of the international consensus governing space activity. From this, there is the danger of diplomatic tension and possibly of armed conflict over the resources and the method by which any profits are shared. There is little doubt that this is a real risk with grave consequences. It is not, however, the inevitable consequence of space mining activity. The discussion will now seek to outline potential legal and policy solutions to mitigate some or all of this risk. A new solution to space property rights could not only keep the peace both on earth and in space, but also provide the certainty to unlock investment and see the emergence of a viable, lucrative space mining industry.

10.7 MITIGATING THE RISK FROM SPACE MINING: A NEW SOLUTION TO PROPERTY RIGHTS?

There is recognition, both by the space mining industry and nation states that space mining requires international legitimacy in order to work. There are a number of reasons for this, not least of which is that space activities do not occur in a political vacuum and are not inseparable from their broader geopolitical context. Commercial entities need an international market place for their resources in order for their industry to be economically viable. This requires that their property rights over the extracted resources have international legitimacy, especially as the so-called "new space revolution" is making the satellite market increasingly international. Indeed, mining companies rank security of property rights as their highest priority when investing in mining ventures (see Dula and Zheniun 2015). Furthermore, international dialogue and cooperation over space resources is needed to avert the potential conflicts discussed above.

10.7.1 Space Resources and International Bodies

Despite the dangers posed by fracturing the international consensus, there is some evidence of a will to cooperate on an international level. The ongoing discussions during the sessions of the COPUOS, the Legal Subcommittee in particular, shows that the subject is getting precisely the level of scrutiny needed. This is particularly important as COPUOS has been the source of all five of the existing space law treaties and is seen as the body as having the most legitimacy with regard to the international governance of peaceful activities in outer space. Additionally, there is The Hague Space Resources Governance Working Group,* a group of legal, policy, and scientific experts as well as industry representatives and other stakeholders who have been discussing a set of Building Blocks for the Development of an International Framework on Space Resource Activities.† It is recognized that the space mining industry is in its infancy and any proposed governance framework, either at the national or international level, needs to keep this in mind. At this stage, it needs to focus on general principles rather than technical details, especially about the actual mining process itself. This will be facilitated by the use of the principle of "adaptive governance" (under Building Block 4.2a) whereby regimes and structures

* http://law.leiden.edu/organisation/publiclaw/iiasl/working-group/the-hague-space-resources-governance-working-group.html.

† The Hague Working Groupe Draft Building Blocks on Space Resource Activities 2017—http://media.leidenuniv.nl/legacy/draft-building-blocks.pdf.

need to be adaptable to deal with developments without having to revise or abandon the overall governance framework.

From these discussions, it would appear that coordination and discussion should be prioritized over excessive harmonization. There may come a time when harmonization is necessary, perhaps via a binding legal instrument like a new treaty, but given the state of the technology and infrastructure it is premature to discuss such things at this stage. There is a need to establish—as a normative value—that it is within the sovereign right of a state to produce national legislation for the "authorization and supervision" of space mining activities conducted by its nationals without being in violation of Article II of the OST. Once this is accepted, it is also clear that states need the right to devise their own methods for doing so, especially if there is little expertise on either the activity or the regulation of the activity. It is suggested that it is sensible for the international community to consider coordination of both activities and national legislation. Coordination of activities is the most pressing concern as there is the prospect that multiple operators from multiple states could be interested in the same celestial body and for the early operators to go after the "low hanging fruit." It is not only in the interest of states to avoid conflict and interference with the operations of other companies and states, it is actually a requirement of the Article IX of the OST (and indirectly through Article III). As for legislative cooperation, the space industry itself is largely and increasingly international in focus, which is also the trend within the space mining industry. Both Planetary Resources and Deep Space Industries, while having headquarters in the United States also have subsidiaries in Luxembourg, and similarly iSpace, while based in Japan also has a subsidiary in Luxembourg.

10.7.2 Space Resources Dispute Resolution

Fundamentally, there is a need for dispute resolution mechanisms. It may be premature to discuss a formal dispute resolution system beyond those already existing for disputes of an international nature. It would, however, be prudent to discuss arrangements before a dispute arises, and some form of "soft law" agreement between states or even perhaps industry guidelines could be of enormous benefit to all concerned. In concert with this, it is also worth taking some time to consider the provisions of Article 11 of the Moon Agreement, which could provide the legal certainty and security the space mining industry desires. The international regime called for in Article 11(5) is not an unreasonable idea. There are no specifications as to the nature of this international regime. There may be a common expectation that the international regime would duplicate the International Seabed Authority, but there is no detail in the Moon Agreement requiring it. Nonbinding, voluntary mechanisms could play an important role in shaping normative solutions based on incentivized participation. Such incentives will be invaluable in overcoming political reticence to commit to the Moon Agreement.

Article 11(7) calls for "an equitable sharing by all States Parties in the benefits derived from those resources…" It is important to note several aspects to this provision. First, it calls for an equitable sharing, not an equal sharing. While equity and equality are often related they do mean different things (Christol 1981). Equity is more concerned with fairness and there is certainly an argument to be made that those who actually undertake the effort and risk to extract resources should be the primary (if not sole) beneficiary. The wording of Article 11(7) is also significant stating that equitable distribution "shall be given special consideration;" this does not require a redistribution of the benefits of resource extraction to all states on Earth. Any advocacy of resource sharing and the creation of expensive, multinational bodies to administer such sharing (outside of the control of sovereign states) would appear to be politically naïve. It is unlikely that, in its present form, any aspect of Article 11 of the Moon Agreement would be palatable to major space actors.

Yet even in this respect there are precedents giving cause for some optimism. The aforementioned UNCLOS, dealing with mining of the seabed is illustrative of potential avenues of consensus and cooperation. UNCLOS was negotiated around the same time as the Moon Agreement and also contains the Common Heritage of Mankind principle (as stated above). Crucially, it too was regarded as being a "failed treaty" as Western states largely avoided it, not desiring to have to give up their technology to lesser-developed states. It was, however amended in 1994, following protracted negotiations and now a majority of states have signed up to UNCLOS with the noticeable exception of the United States (Adolph 2006). Admittedly, it is unlikely the Moon Agreement could be revived in its current form. The agreement is toxic politically, particularly in the United States, and discussions attempting to revitalize the Moon Agreement remain moribund. Despite this, it is contended that the essence of the Moon Agreement, particularly Article 11 should not be discounted for the reasons listed above. Indeed, a form of words similar to Article 11 of the Moon Agreement could, at some point form the basis of an internationally legitimate regime.

10.8 CONCLUSION

As stated at the outset, this chapter has examined the prosaic and administrative risks to space mining. The discussion on the technical environment has illustrated that there are still too many imponderables, in terms of infrastructure and industrial practices, to effectively draw up any meaningful governance framework. The incomplete scientific picture in respect of the actual abundance of resources on asteroids and NEOs compounds this lack of certainty. Where the drafters of the treaty in 1967 based their deliberations on the practices of the state parties at the time, such contemporary activity is not available for those looking to legislate on space mining at the moment. Given the incomplete picture, augmented only with supposition and educated guesswork, the temptation is to recommend lawmakers adopt a cautious wait-and-see approach rather than enshrining a system possibly not reflecting the reality of space mining operations.

The picture for cooperation is not without cause for optimism. The space law community, both within COPUOS and other informal groupings, has responded with alacrity. While there has been some posturing, there still exists a dialogue over ways in which resource ownership questions and Article II can be addressed. The very fact that discussions are ongoing is a source of optimism. A wholesale renegotiation of the 1967 treaty appears unlikely in the current geopolitical environment. The Moon Agreement of 1979 at first sight appears to be an unlikely savior, but there is much in it already accepted as normative. The evolution of UNCLOS suggests that even the most moribund of international agreements can be revived if there is sufficient political will. Nonetheless, only the most optimistic of advocates of the Moon Agreement would overlook the difficulties in garnering widespread support for a treaty widely regarded as having failed due to the uncertainties of the CHM principle.

A collaborative approach at this stage appears to have limited chances of success. States such as the United States and Luxembourg have already shown a willingness to fill the legal and policy void around space resources with bespoke national provisions designed to embolden investors and energize companies into developing space mining as a serious endeavor. These domestic laws eschew the difficulties of the appropriation aspect of Article II of the OST by focusing on the licensing and supervisory duties incumbent under Article VI, and by seeing mining as an element of usage permitted under Article I. The genie of national legislation is out of the bottle and the risk is there will be a fracturing along the fault lines of Article II. Such a schism could significantly undermine the peaceful and largely cooperative nature of space governance existing since the signing of the OST. Independent mining and the perception of a colonialist approach to space mining could further exacerbate diplomatic tensions.

Space mining may be the start of the divergence of space law. This could see a move from a uniform framework under international law to a more fragmented situation where national law is the primary instrument—this process may already be underway. The future of the international framework may depend on whether other states decide to enact legislation providing for national property rights or whether an enduring consensus can be found along the lines suggested. It is a legitimate position to question whether the international approach to space governance is the right way to go, given the changing geopolitical environment. The risk in the OST regime collapsing is that it forfeits stability and the acceptance in free and unfettered access to space. A new, internationally recognized space property rights framework—agreed on by all stakeholders—must surely be the aim of ongoing discussions. Even if that framework is predicated on broadly accepted national laws, the risk of doing nothing far outweighs the risk of being pro-active. The creation of such a framework would facilitate exploration, encourage investment, and pave the way for the creation of an infrastructure, which could see humanity emerge as a truly interstellar species.

REFERENCES

Adolph, J., 2006. The recent boom in private space development and the necessity of an international framework embracing private property rights to encourage investment. *The International Lawyer* 40: 961–985.

Bilder, R.J., 2009–10. A legal regime for the mining of helium-3 on the Moon: U.S. policy options. *Fordham International Law Journal* 33: 243–299.

Blount, P., 2012. Renovating space: The future of international space law. *Denver Journal of International Law & Policy* 40: 515–532.

Christol, C., 1981. Evolution of the Common Heritage of Mankind Principle. *Western State University International Law Journal* 1: 63–74.

Cookson, C., 2017. Space Mining Takes Giant Leap from Sci-Fi to Reality. *London: Financial Times*, October 19, 2017. Available at: https://www.ft.com/content/78e8cc84-7076-11e7-93ff-99f383b09ff9.

David, L., 2015. Is Moon mining economically feasible? *Space.com* January 7, 2015. Available at: https://www.space.com/28189-moon-mining-economic-feasibility.html.

David, L., 2016. Inside ULA's plan to have 1000 people working in space by 2045. *Space.com* June 29, 2016. Available at: https://www.space.com/33297-satellite-refueling-business-proposal-ula.html.

Di Martino, M., Carbognani, A., De Sanctis, G., Zappala, V., and Somma, R., 2009. *The Asteroid Hazard: Evaluating and Avoiding the Threat of Asteroid Impacts*. European Space Agency.

Dula, A.M. and Zhang Zheniun, eds., 2015. *Space Mineral Resources: A Global Assessment of the Challenges and Opportunities*. International Academy of Astronautics.

Elvis, M., 2013. Prospecting asteroid resources. In *Asteroids: Prospective Energy and Material Resources*. ed. V. Badescu, 81–129. London, Springer.

Foust, J., 2012. Planetary resources believes asteroid mining has come of age. *The Space Review* April 30, 2012. Available at: http://www.thespacereview.com/article/2074/1.

Freeland, S. and Jakhu, R., 2009. Article II. In *Cologne Commentary on Space Law*, vol. 1. eds. S. Hobe, B. Schmidt-Tedd and K. Schrogl, 221–271. Cologne: BWV Verlag.

Gangale, T., 2009. *The Development of Outer Space: Sovereignty and Property Rights in International Space Law*. New York: Praeger.

Granvick, M., et al., 2013. Earth's temporarily-captured natural satellites—The first steps towards utilization of asteroid resources. In *Asteroids: Prospective Energy and Material Resources*, ed. V. Badescu, 151–167. London, Springer.

Henry, C., 2017. *Airbus to challenge SSL, Orbital ATK with new space tug business. Space News* September 28, 2017. Available at: http://spacenews.com/airbus-to-challenge-ssl-orbital-atk-with-new-space-tug-business/?utm_content=buffer46444&utm_medium=social&utm_source=twitter.com&utm_campaign=buffer.

Hobe, S., 2009. Article I. In *Cologne Commentary on Space Law*, vol. 1. eds S. Hobe, B. Schmidt-Tedd and K. Schrogl, 25–43. Cologne: BWV Verlag.

Jakhu, R., 2005. Twenty years of the moon agreement: Space law challenges for returning to the moon. *ZLW* 2005: 243–260.

Jakhu, R., Freeland, S., Hobe, S., and Tronchetti, F., 2013. Article 11 (Common Heritage of Mankind/International Regime). In *Cologne Commentary on Space Law*, vol. 2. eds. S. Hobe, B. Schmidt-Tedd and K. Schrogl, 388–399. Cologne: BWV Verlag.

Johnson-Freese, J., 2017. *Space Warfare in the 21st Century: Arming the Heavens*. New York: Routledge.

Larsen, P.B., 2014. Asteroid legal regime: Time for a change? *Journal of Space Law* 39: 275.

Lewis, J.S., 1997. *Mining the Sky: Untold Riches from the Asteroids, Comets and Planets*. New York: Helix Books.

Lewis, J.S., 1999. Tapping the Waters of Space' (Spring 1999) 10 Scientific American Presents 100, 100–103.

Lewis, J.S., 2015. *Asteroid Mining 101: Wealth for the New Space Economy*. Deep Space Industries.

Listner, M.J., 2014. Redux: It's time to rethink international space law. *The Space Review*: November 24 2014. Available at: http://www.thespacereview.com/article/2647/1.

Lyall, F. and Larsen, P.B., 2009. *Space Law: A Treatise*. London: Ashgate.

Newman, C.J., 2015. Seeking tranquillity: Towards an optimal solution for lunar regulation. *Space Policy* 33: 29–37.
Oduntan, G., 2005. Imagine there are no possessions: Legal and moral basis of the common heritage principle in space law. *Manchester Journal of International Economic Law* 2: 30–58.
Pop, V., 2009. *Who Owns the Moon? Extra-terrestrial Aspects of Land and Mineral Resources Ownership.* London: Springer.
Pop, V., 2013. Legal considerations on asteroid exploitation and deflection. In *Asteroids: Prospective Energy and Material Resources.* ed. V. Badescu, 659–680. London, Springer.
Ryan, M.H. and Kutschera, I., 2013. The case for asteroids: Commercial concerns and considerations. In *Asteroids: Prospective Energy and Material Resources.* ed. V. Badescu, 646–647, London: Springer.
Shepard, M.K., 2015. *Asteroids: Relics of Ancient Time.* Cambridge: Cambridge University Press.
Tronchetti, F., 2015a. Legal aspects of space resource utilization. In *Handbook of Space Law.* eds. F. von der Dunk and F. Tronchetti, 769–813. London, Edward Elgar.
Tronchetti, F., 2015b. The space resource exploration and utilization act: A move forward or a step back? *Space Policy* 34: 6–9.
UNCOPUOS, 2016. *Report of the Legal Subcommittee on its Fifty-Fifth Session* (Vienna April 27, 2016) UN Doc A/AC.105/1113.
von der Dunk, F., 2012. Contradictio in terminis or Realpolitik? A qualified plea for a role of "soft law" in the context of space activities. In *Soft Law in Outer Space: The Function of Non-binding Norms in International Space Law,* ed. Irmgard Marboe. 31–56, BWV Verlag.
Willis, J., 2016. *All These Worlds Are Yours: The Scientific Search for Alien Life.* London: Yale University Press.
Woerner, J., 2016. *Moon Village: A vision for global cooperation and space 4.0.* ESA. Available at: http://www.esa.int/About_Us/Ministerial_Council_2016/Moon_Village.
Zullo, K.M., 2001-2002. The need to clarify the status of property rights in international space law. 90 Georgetown Law Journal 2413–2444.

Index

Note: Page numbers followed by "*fn*" indicate footnotes

A

D

E

F

G

H

N

R

For Product Safety Concerns and Information please contact our EU representative GPSR@taylorandfrancis.com Taylor & Francis Verlag GmbH, Kaufingerstraße 24, 80331 München, Germany

Batch number: 10726532

Printed by Printforce, the Netherlands